Planning for Seismic Rehabilitation: Societal Issues

Issued by FEMA in furtherance of the
Decade for Natural Disaster Reduction.

BUILDING SEISMIC SAFETY COUNCIL

Of the National Institute of Building Sciences

Program
on
Improved
Seismic
Safety
Provisions

PLANNING FOR SEISMIC REHABILITATION: SOCIETAL ISSUES

FEMA 275

The **Building Seismic Safety Council** (BSSC) was established in 1979 under the auspices of the National Institute of Building Sciences as an entirely new type of instrument for dealing with the complex regulatory, technical, social, and economic issues involved in developing and promulgating building earthquake hazard mitigation regulatory provisions that are national in scope. By bringing together in the BSSC all of the needed expertise and all relevant public and private interests, it was believed that issues related to the seismic safety of the built environment could be resolved and jurisdictional problems overcome through authoritative guidance and assistance backed by a broad consensus.

The BSSC is an independent, voluntary membership body representing a wide variety of building community interests. Its fundamental purpose is to enhance public safety by providing a national forum that fosters improved seismic safety provisions for use by the building community in the planning, design, construction, regulation, and utilization of buildings.

To fulfill its purpose, the BSSC: (1) promotes the development of seismic safety provisions suitable for use throughout the United States; (2) recommends, encourages, and promotes the adoption of appropriate seismic safety provisions in voluntary standards and model codes; (3) assesses progress in the implementation of such provisions by federal, state, and local regulatory and construction agencies; (4) identifies opportunities for improving seismic safety regulations and practices and encourages public and private organizations to effect such improvements; (5) promotes the development of training and educational courses and materials for use by design professionals, builders, building regulatory officials, elected officials, industry representatives, other members of the building community, and the public; (6) advises government bodies on their programs of research, development, and implementation; and (7) periodically reviews and evaluates research findings, practices, and experience and makes recommendations for incorporation into seismic design practices.

BOARD OF DIRECTION: 1997

Chairman	Eugene Zeller, City of Long Beach, California
Vice Chairman	William W. Stewart, Stewart-Scholberg Architects, Clayton, Missouri (representing the American Institute of Architects)
Secretary	Mark B. Hogan, National Concrete Masonry Association, Herndon, Virginia
Ex-Officio	James E. Beavers, James E. Beavers Consultants, Oak Ridge, Tennessee
Members	Eugene Cole, Carmichael, California (representing the Structural Engineers Association of California); S. K. Ghosh, Portland Cement Association, Skokie, Illinois; Nestor Iwankiw, American Institute of Steel Construction, Chicago, Illinois; Gerald H. Jones, Kansas City, Missouri (representing the National Institute of Building Sciences); Joseph Nicoletti, URS/John A. Blume and Associates, San Francisco, California (representing the Earthquake Engineering Research Institute); Jack Prosek, Turner Construction Company, San Francisco, California (representing the Associated General Contractors of America); W. Lee Shoemaker, Metal Building Manufacturers Association, Cleveland, Ohio; John C. Theiss, Theiss Engineers, Inc., St. Louis, Missouri (representing the American Society of Civil Engineers); Charles Thornton, Thornton-Tomasetti Engineers, New York, New York (representing the Applied Technology Council); David P. Tyree, American Forest and Paper Association, Colorado Springs, Colorado; David Wismer, Department of Licenses and Inspections, Philadelphia, Pennsylvania (representing the Building Officials and Code Administrators International); Richard Wright, National Institute of Standards and Technology, Gaithersburg, Maryland (representing the Interagency Committee for Seismic Safety in Construction)
BSSC Staff	James R. Smith, Executive Director; Thomas Hollenbach, Deputy Executive; Lee Lawrence Anderson, Director, Special Projects; Claret M. Heider, Program Manager/Technical Writer-Editor; Mary Marshall, Administrative Assistant; Patricia Blasi, Administrative Assistant

PLANNING FOR SEISMIC REHABILITATION: SOCIETAL ISSUES

FEMA 275

Developed for the Building Seismic Safety Council
by ROA (Robert Olson Associates, Inc.)
with the support of the
Federal Emergency Management Agency

BUILDING SEISMIC SAFETY COUNCIL
of the National Institute of Building Sciences
Washington, D.C.
1998

This report was prepared under Cooperative Agreement EMW-91-K-3602 between the Federal Emergency Management Agency and the National Institute of Building Sciences.

Building Seismic Safety Council activities and products are described at the end of this report. For further information, contact the Building Seismic Safety Council, 1090 Vermont, Avenue, N.W., Suite 700, Washington, D.C. 20005; phone 202-289-7800; fax 202-289-1092; e-mail bssc@nibs.org. Copies of this report may be obtained by contacting the FEMA Publication Distribution Facility at 1-800-480-2520.

FOREWORD

In 1984, the Federal Emergency Management Agency (FEMA) initiated a comprehensive, and closely coordinated program to develop a body of knowledge in support of building practices that would increase the ability of existing buildings to withstand the forces of earthquakes. Societal issues inherent in seismic rehabilitation processes also have received attention. At a cumulative cost of about $26 million, this FEMA effort has generated two dozen publications and a number of software programs and audio-visual training materials for use by design professionals, building regulatory personnel, educators, researchers, and the general public. The program has proceeded along separate but parallel approaches in dealing with both private sector and federal buildings.

Already available from FEMA to private sector practitioners and other interested parties is a "technical platform" of consensus criteria on how to deal with some of the major engineering aspects of the seismic rehabilitation of buildings. Completed in 1992, this technical material comprises a trilogy with supporting documentation: a method for the rapid identification of buildings that might be hazardous in an earthquake and which can be conducted without gaining access to the buildings themselves; a methodology for a more detailed evaluation of a building that identifies structural flaws that have caused collapse in past earthquakes and might do so again in future earthquakes, and a compendium of the most commonly used techniques of seismic rehabilitation.

Along with this volume, the culminating activity in the field of seismic rehabilitation is the completion of a comprehensive set of nationally applicable guidelines with commentary on how to rehabilitate buildings so that they will better withstand earthquakes. Known as the *NEHRP Guidelines for the Seismic Rehabilitation of Buildings* (FEMA 273) and the *Commentary on the Guidelines for the Seismic Rehabilitation of Buildings* (FEMA 274), these volumes, the results of a multiyear, multimillion dollar effort, represent a first of its kind in the United States. The *Guidelines* allow practitioners to choose design approaches consistent with different levels of seismic safety as required by geographic location, performance objective, type of building, use or occupancy, or other relevant considerations. The *Guidelines* documents also include analytical techniques that will assist in generating reliable estimates of the expected earthquake performance of rehabilitated buildings. This extensive platform of materials fills a significant gap in that portion of the National Earthquake Hazards Reduction Program (NEHRP) focusing on the seismic safety of existing buildings.

The *Guidelines* documents were given consensus review by representatives of a broad spectrum of users including the construction industry; building designers; building regulatory organizations; building owners and occupant groups; academic and research institutions; financial establishments; local, state, and federal levels of government; and the general public. This process helped to ensure the national applicability of the *Guidelines* documents and encourage widespread acceptance and use by practitioners. It is expected that, with time, the *Guidelines* will be referenced or adapted by standards-setting groups and model building code organizations and will thereby diffuse widely into building practices across the United States.

This volume complements the technical materials principally oriented to design professionals in the *Guidelines* documents. Because of the complexities and possible disruption caused by seismic rehabilitation projects, this volume's title, *Planning for Seismic Rehabilitation: Societal Issues*, calls attention to two important themes: that careful planning can minimize possibly difficult societal problems and that there exists a wide range of societal issues that may be more significant in rehabilitation projects than in new construction. In many ways, this publication is intended to provide a "heads up" to those who are considering individual or multiple building, construction class or use, or area-focused seismic rehabilitation efforts.

This volume exploring societal issues reflects very generous contributions of time and expertise on the part of many individuals, contributions that are warmly acknowledged. FEMA is particularly grateful for the efforts of the BSSC and its consultant Robert Olson, the Project Oversight Committee, and the BSSC Project Committee and Seismic Rehabilitation Advisory Panel.

Federal Emergency Management Agency

PREFACE and ACKNOWLEDGMENTS

In August 1991, the National Institute of Building Sciences (NIBS) entered into a cooperative agreement with the Federal Emergency Management Agency (FEMA) for conduct of a comprehensive seven-year program leading to the development of a set of nationally applicable guidelines for the seismic rehabilitation of existing buildings. Under this agreement, the Building Seismic Safety Council (BSSC) served as program manager with the American Society of Civil Engineers (ASCE) and the Applied Technology Council (ATC) working as subcontractors. Initially, FEMA provided funding for a program definition activity designed to generate the detailed work plan for the overall program. The work plan was completed in April 1992 and in September FEMA contracted with NIBS for the remainder of the effort.

The major objectives of the project were to develop a set of technically sound, nationally applicable guidelines (with commentary) for the seismic rehabilitation of buildings; to achieve building community consensus regarding the guidelines; and to structure the basis of a plan for stimulating widespread acceptance and application of the guidelines. The technical guidelines documents produced as a result of this project—the *NEHRP Guidelines for the Seismic Rehabilitation of Buildings* (FEMA 273) and its *Commentary* (FEMA 274)—are intended to serve as a primary resource on the seismic rehabilitation of buildings for the use of design professionals, educators, model code and standards organizations, and state and local building regulatory personnel.

As noted above, the project work involved the ASCE and ATC as subcontractors as well as groups of volunteer experts and paid consultants, and it was structured to ensure that the technical guidelines writing effort benefited from consideration of: the results of completed and ongoing technical efforts and research activities; societal issues, public policy concerns, and the recommendations presented in an earlier FEMA-funded report on issues identification and resolution; cost data on application of rehabilitation procedures; the reactions of potential users; and consensus review by a broad spectrum of building community interests.

While overall management has been the responsibility of the BSSC, responsibility for conduct of the specific project tasks was shared by the BSSC with ASCE and ATC. Specific BSSC tasks were completed under the guidance of a BSSC Project Committee. To ensure project continuity and direction, a Project Oversight Committee (POC) was responsible to the BSSC Board of Direction for accomplishment of the project objectives and the conduct of project tasks. Further, a Seismic Rehabilitation Advisory Panel reviewed project products as they developed and advised the POC on the approach being taken, problems arising or anticipated, and progress made. Three user workshops also were held during the course of the project to expose the project and various drafts of the *Guidelines* documents to review by potential users of the ultimate project product.

The final drafts of the *Guidelines* and its *Commentary* were submitted to the BSSC member organizations for balloting in October-December 1996 and June-July 1997. The final versions of the consensus-approved documents were transmitted to FEMA for publication in September 1997.

This document was developed for the Building Seismic Safety Council by ROA (Robert Olson Associates, Inc.) to serve as an additional resource to provide those considering seismic rehabilitation with insights into the complex economic, social, and political issues surrounding such efforts. The BSSC is grateful to Mr. Olson for sharing his professional expertise and participating throughout the project.

The BSSC also wishes to acknowledge the wide variety of groups that provided Mr. Olson with helpful contributions and suggestions. Special appreciation is extended to the members of the BSSC Project Committee and Seismic Rehabilitation Advisory Panel, the participants in the users' workshops held during the *Guidelines* development effort, and the Advisory Committee on Social and Policy Issues formed for this project by the Earthquake Engineering Research Institute—all of whom provided valuable advice and comments (see Appendix B for committee/panel membership lists). The BSSC also wishes to acknowledge the efforts of Ugo Morelli, FEMA Project Officer, and his technical advisor, Diana Todd, both of whom provided thoughtful and constructive suggestions during that have immeasurably improved the products of the project.

It should be noted that recommendations resulting from the concept work of the BSSC Project Committee have resulted in initiation of a case studies project that will focus on the development of seismic rehabilitation designs for over 40 buildings selected from an inventory of buildings determined to be seismically deficient under the implementation program of Executive Order 12941 and determined to be considered

"typical of existing structures located throughout the nation."

Feedback from those reading this *Societal Issues* volume and using the *Guidelines* documents outside the case studies project is strongly encouraged. Further, the curriculum for a series of education/training seminars on the *Guidelines* is being developed and a number of seminars are scheduled for conduct in 1998. Those who wish to provide feedback or with a desire for information concerning the seminars should direct their correspondence to: BSSC, 1090 Vermont Avenue, N.W., Suite 700, Washington, D.C. 20005; phone 202-289-7800; fax 202-289-1092; e-mail bssc@nibs.org.

Eugene Zeller, BSSC Chair

EXECUTIVE SUMMARY

Those involved in the complex process of preparing the *NEHRP Guidelines for the Seismic Rehabilitation of Buildings* and its *Commentary* (referred to in this publication as the *Guidelines* or the *Guidelines* documents) recognized from the outset the importance of helping users deal with the social, economic, and public policy complexities of rehabilitation. Indeed, the Executive Director of the Building Seismic Safety Council, the managing organization for this project, noted that seismic rehabilitation decision-makers "possibly are not technically oriented but will have to say yea or nay on incorporating information from the *Guidelines* into local practices, be they business or regulatory."

This *Societal Issues* volume has been prepared to acquaint potential users of the *Guidelines* documents with typical problems unrelated to design and construction processes that might arise when planning or engaging in seismic rehabilitation projects and programs. Further, it is intended to alert readers to the difficulties inherent in implementing seismic rehabilitation recommendations.

The goals of seismic rehabilitation are important. They include, above all, protecting life and property in future earthquakes as well as protecting investments, lengthening a building's usable life, reducing demands on post-earthquake search and rescue resources, protecting historic structures, shortening business interruption time, maintaining inventories and customers, and reducing relocation needs/demands. Other worthy goals include limiting the need for post-earthquake emergency shelter and temporary housing, minimizing the release of hazardous substances, conserving natural resources, avoiding the costly processes of settling insurance claims and applying for post-disaster aid, protecting savings and contingency funds, reducing the amount of debris to be removed, and facilitating an earthquake-stricken community's return to normal patterns of activity.

This publication is structured to emphasize two basic user-oriented concepts. The first is a four- step iterative process that outlines a set of decision points so the user can determine whether seismic rehabilitation efforts are needed and, if so, their potential scope. The second offers a simple "escalation ladder" to help users understand the degree of conflict inherent in and the implications of choosing what, if any, seismic rehabilitation strategies to follow.

The four-step decision process includes:

- Defining the problem by conducting preliminary and, if needed, detailed analyses of the risk;

- Developing and refining the alternatives for addressing seismic rehabilitation;

- Adopting an approach and an implementation strategy; and

- Securing the needed resources and implementing the seismic rehabilitation measures.

The strategies available to those who become involved with seismic rehabilitation will reflect the mixture of private efforts and governing public policies existing in the specific context (e.g., a city). Attrition is one choice and has the least conflict. A second choice is purely voluntary rehabilitation, but even this approach may engender some conflict as government becomes involved in the permitting process. The third choice involves a more proactive role of government and, therefore, a potentially higher level of conflict; it entails informally encouraging owners to rehabilitate their buildings by establishing some standards and triggers and then negotiating the scope of work on a case-by-case basis as a condition of being granted the necessary permits. The fourth and final strategic choice and the one with the highest degree of conflict centers on government mandation of seismic rehabilitation—i.e., the establishment of seismic rehabilitation ordinances defining which types or uses of buildings require rehabilitation, the applicable standards, reporting and inspection requirements, time frames for compliance, and penalties for not doing so.

In recognition of the fact that each building is unique, this publication also examines the wide spectrum of socioeconomic issues that may face those involved in seismic rehabilitation efforts. Each is discussed in terms of the nature of the problem, typical issues, and some example solutions. Considered are problems related to historic properties, the distribution of economic impacts, occupant dislocation, business interruption, effects on the housing stock, rehabilitation triggers, financing rehabilitation, legal concerns, and selection of rehabilitation targets.

Inasmuch as the intended users of the *Guidelines* documents and this publication are most likely to be local building and planning officials, private owners and consulting design professionals, three illustrative "application scenarios" are presented. Each scenario presents a situation (for a private company facilities manager; a local government city manager and building official; and a consulting engineer) and a list of considerations that would commonly have to be addressed.

The economic, social, and political complexities and the varying seismic environments of the United States are such that seismic rehabilitation programs will have to be tailored to thousands of individual situations. This publication therefore provides an extensive reference section to help the reader locate additional applicable materials.

CONTENTS

Chapter 1
WHY SEISMIC REHABILITATION?

WHY REHABILITATION?

The core argument for the seismic rehabilitation of buildings is that rehabilitated buildings will provide increased protection of life and property in future earthquakes, thereby resulting in fewer casualties and less damage than would otherwise be the case. It is a classic mitigation strategy not unlike preventive medicine. On the human level, more earthquake-resistant buildings will mean fewer deaths and injuries in an event and therefore lower demand on emergency medical services, urban search and rescue teams, fire and law enforcement personnel, utilities, and the providers of emergency shelter. In the commercial sector, less damage to structures will mean enhanced business survival and continued ability to serve customers and maintain markets or market shares. More specifically, for commercial enterprises seismic rehabilitation will better protect physical and financial assets; reduce inventory loss; shorten the business interruption period; avoid the need for relocation; and minimize secondary effects on suppliers, shippers, and other businesses involved in support services or product cycles. For governments, less damage to government structures will mean continued services and normal processes or at least minimal interruptions. If government structures come through an earthquake with little or no damage, agencies will not have to relocate services, and public officials can respond to the immediate and long-term demands placed on them by the event. In short, seismic rehabilitation as a pre-event mitigation strategy actually will improve post-event response by lessening life loss, injury, damage, and disruption.

Seismic rehabilitation also will help achieve other important goals, that contribute to business and community well-being. For example, seismic rehabilitation will::

- Reduce community economic and social impacts (e.g., less loss of employment and increased blighted areas resulting from an earthquake and less loss of tax revenues to support public services).

- Minimize the need for and the process management time required to obtain disaster assistance as well as the financial impacts of filing insurance or disaster assistance claims, seeking loans or grants, and liquidating savings or contingent reserves.

- Help to protect historic buildings, structures, or areas that represent unique community values and that provide the residents with a sense of their unique histories.

- Minimize impacts on such critical community services as hospitals and medical care facilities, whether or not such services are provided by private, nonprofit, or government entities.

- Support the community's post-earthquake need to return to a pattern of normal activities by helping to ensure the early reopening of business and civic facilities (e.g., functioning schools, stores, and government offices). In addition to reducing demands for immediate assistance, such as providing emergency shelter and food, restoring normal activities as soon as possible contributes greatly to the psychological well-being of a community – e.g., children return to school, parents return to work, businesses reopen, and links with the broader "outside world" are restored.

- Minimize the many and often subtle direct and indirect socioeconomic impacts of earthquakes, some of which emerge slowly but often last a long time. For example, after a disaster, low-income residents often become displaced which adds to any existing homeless problem and increases the burden on community services and charitable organizations, often reducing their abilities to provide regular services. Further, marginal businesses may not be able to reopen, thus weakening a community's economic and social fabric and reducing tax revenues, which may result in a shift in the tax structure to pay for public services. Finally, the distribution of impacts may mean that adjacent areas gain at the expense of the damaged areas.

- Reduce the difficult environmental impacts of earthquakes. These include, for example, the need to dispose of large quantities of debris, the release of asbestos in damaged buildings, and the contamination of the air and water with spilled hazardous materials.

In sum, the rehabilitation of existing buildings to better resist future damaging earthquakes truly is "preventive medicine." While seismic rehabilitation costs money, it can significantly reduce future losses and, in economic terms, can be considered an investment to protect assets currently at risk. Emergency response capabilities, as good as they are in U.S. communities, are no substitute for amelioration of the direct and indirect losses to each citizen's physical assets and each community's infrastructure.

WHAT FOLLOWS?

Completing this *Societal Issues* volume are five additional chapters plus an appendix to help the reader achieve the multiple goals of seismic rehabilitation.

Chapter 2 provides a decision-making guide to support the analysis and implementation of efforts to seismically strengthen buildings. Chapter 3 describes the broad context in which seismic rehabilitation occurs, explains how different approaches involve various complexities and degrees of conflict, and provides guidance and case study examples of various approaches and tactics to achieve seismic rehabilitation. Chapter 4 examines a wide range of typical societal problems and explores various ways of addressing them. Chapter 5 presents three application scenarios designed to help the user understand his or her situation and the factors that may be involved in initiating a seismic rehabilitation effort. Chapter 6 points the reader toward some of the socioeconomic literature related to seismic rehabilitation while the Appendix provides a detailed discussion of the four-step process for solving problems. The report concludes with an overview of the purpose and activities of the Building Seismic Safety Council and a list of those involved in the *Guidelines* project.

Chapter 2
A DECISION-MAKING GUIDE

INTRODUCTION

While the seismic rehabilitation of existing buildings presents many of the same challenges to private as well as public sector decision-makers, this publication is intended primarily for local government officials, especially those in planning, redevelopment and building departments, and public agency and private engineers who find themselves involved in the public policy aspects of seismic rehabilitation.

Despite the fact that each building has "its own story" when it comes to seismic rehabilitation, similar public policy issues reappear so often that providing a generalized approach to achieving seismic rehabilitation is possible. Therefore, a generic, four-step process is outlined for use primarily by local government officials as well as, building owners, engineers, and/or private consultants seeking approval from local governments to seismically rehabilitate a building or group of buildings.

Secondarily, this publication is directed toward private-sector decision-makers. The term "private sector" is admittedly quite broad, encompassing the owner of one office building in a small city in a low seismic risk (and awareness) zone, the owner of multiple-unit apartment buildings in a zone of moderate risk (and awareness), a large corporation with facilities in high seismic risk (and awareness) zones, and all those in between.

Nonetheless, despite obviously different contexts and specific problems, the shared nature of the earthquake-vulnerable structure problem establishes certain commonalities between the private and public sectors. Although some parts of this publication may be more relevant than others, the hope is that it will be useful to corporate facility managers who wish to seismically rehabilitate a building or group of buildings and must secure appropriate approvals and support from chief executive officers, boards of directors, or clients. It is important to note, however, that the engineering expertise of a design professional (architect, engineer, code official) is a prerequisite to the appropriate use of the *Guidelines* documents.

It should be noted that even if community or private-sector decision-makers responsible for one or more types of earthquake-vulnerable structures anticipate and address the social, economic, and political complications inherent in seismic rehabilitation, the problems will not be eliminated. This approach will, however, facilitate their management. In addition, effectively managing the human or nontechnical problems of seismic rehabilitation hopefully will make the use of the separate but companion engineering publications, the *Guidelines* documents, more tailored and therefore more sensitive to particular situations and environments.

AN OVERVIEW OF THE FOUR-STEP PROCESS

A common four-step problem-solving process follows:

1. Defining the problem

   1A. Conducting preliminary analysis

   1B. Conducting detailed analysis (+ feedback)

2. Developing and refining alternatives (+ feedback)

3. Adopting an approach and implementation strategy (+ feedback)

4. Securing resources and implementing (+ feedback)

As in many processes of this type, this generic four-step model emphasizes the feedback function at every step because no existing building seismic rehabilitation effort can possibly succeed in isolation, no matter how splendid the technical components. Seismic rehabilitation takes place in a wide variety of socioeconomic and political contexts, and continuous feedback and adjustments are necessary for success. The number of affected buildings, the acceptable level of risk defined by the selected rehabilitation performance objectives, the duration of the program, the cost, and the social and economic impacts are interdependent. By the very number and nature of the variables, seismic rehabilitation decision-making is very complex for it must balance so many considerations.

The level of detail, amount of data collected, degree of analysis, formality of procedures, and resources committed will vary with the intended use of the engineering publications (the *Guidelines* documents) and with the conditions and circumstances faced by the reader. As a result, given differing community, jurisdictional or corporate contexts, each reader must determine the extent of data collection and analysis of alternatives needed. In other words, each step constitutes a kind of progressive discovery leading to a better understanding of the issues. Each step tests whether the seismic risk justifies the cost and effort involved in taking the next step. Thus, the process is essentially iterative with the steps building on assumptions and estimates of the nature and scope of potential problems and then allowing expansion and refinement of the approach.

Step 1, "Defining the Problem," actually comprises two substeps: "preliminary analysis"and "detailed analysis." Preliminary analysis (Step 1A) entails an initial and perhaps even cursory survey of the general issues raised by an identified earthquake threat. Because earthquake-induced life and property losses tend to be concentrated in building types already known to be vulnerable, once a relatively specific degree of seismic risk and likely consequences have been identified, the issue of seismic rehabilitation arises almost immediately. Therefore, the product of Step 1A is simply a good enough understanding of the seismic risk, the possible scope of potential building rehabilitation efforts, and the implications of such rehabilitation for owners, occupants, and the community so that an informed decision to proceed or not proceed can be made. If a decision is made to proceed, Step 1B, detailed analysis, defines more precisely the nature of the risk and the problem through:

1. Collection of data on the physical nature and policy implications of possible target buildings

2. Refinement and expansion of the initial understanding,

3. Definition of the specific problems and impacts, and

4. Identification of the people and organizations potentially affected by rehabilitation.

The product of Step 1B is a decision to proceed or not proceed given consideration of alternatives and the impact of the decision.

Step 2, "Develop and Refine Alternatives," involves using the data assembled under Step 1B to develop and refine alternative approaches that address the seismic rehabilitation of existing buildings in light of the risk, the costs, and the social and economic impacts. Thus, Step 2 provides a kind of "menu" delineating seismic rehabilitation options for communities in various risk situations. Step 2 usually is a very long and involved process, but the key variables always are the desired performance levels, the scope of the approach, and an estimate of the costs. The first determines how much rehabilitation needs to be accomplished; the second determines how many buildings of what type and use are to be subject to rehabilitation; and the third estimates the cost of each alternative. The outcome of Step 2 is a recommendation, usually from a facilities manager or building official, to the next-level decision-maker(s) on a particular approach to seismic rehabilitation. For public entities, an environmental impact report may be required as part of this step.

Step 3, "Adopt an Approach and Implementation Strategy," is the decision point at which the city or county council, chief executive officer, board, building owner, agency director, or whoever is charged with the final responsibility considers the rehabilitation recommendation, receives input from other sources, and weighs the alternatives (not to be ignored is the alternative of doing nothing). Fundamentally, the decision to act on, modify, or reject a seismic rehabilitation plan is a political decision, whether made by government or a private-sector body. It is a decision that allocates scarce resources, costs, and benefits. It determines who benefits, who pays how much and when, and who bears the indirect costs (e.g., employees, tenants, suppliers,). Finally, the decision to act sets in motion the necessary organizational routines to actually yield activity, in this case seismic rehabilitation.

Step 4, "Secure Resources and Implement," is the critical process that turns a decision to rehabilitate into its physical result--safer, more seismically resistant buildings. Without resources (personnel, budget) to carry out seismic rehabilitation, the adoption of an approach is simply "a piece of paper." In addition, even when the necessary resources are allocated, implementation may be quite extended depending upon the number of buildings slated for rehabilitation, and feedback is perhaps more important here than in any other step. Whoever is charged with overseeing the seismic rehabilitation must be kept apprized of any new techniques or standards that might alter the approach. In addition, the program manager must provide for quality control and must monitor and mitigate, to the extent possible, both the anticipated and the unanticipated socioeconomic and political side effects of seismically rehabilitating buildings.

Chapter 3
SEISMIC REHABILITATION IN CONTEXT

EACH BUILDING HAS ITS OWN STORY

Earthquake-vulnerable buildings exist nationwide, but the earthquake hazard is not uniform across the country. Moreover, awareness of th earthquake hazard, the precursor to any action, varies even more than the hazard itself. Therefore, tackling the earthquake-vulnerable building problem takes place in an incredibly diverse set of geographic, social, economic, and political environments. Further complicating the situation is the fact that no two buildings (even within the same jurisdiction) ever seem to present exactly the same problems. Each building has its own earthquake-vulnerability profile — location, architecture, structural system, occupancy, economic role, and financing. In other words, each building has its own story.

In sum, while few would quibble with the general legitimacy of a policy whose goal is the seismic rehabilitation of earthquake-vulnerable buildings, seismic rehabilitation will be achieved on a city-by-city and, actually, on a building-by-building basis. Such is life in a continent-sized nation with a federal governmental system. The intent of this chapter is to place and explain seismic rehabilitation in various socioeconomic and political contexts and to offer a set of approaches or "models" to inform and guide action.

LOOK BEFORE REHABILITATING

In point of fact, if you are reading this document, you most likely are already beyond what is known in policy analysis as the "problem recognition stage." Precisely because you are reading this volume and presumably the *Guidelines* documents, you are aware of buildings that may be seismically unsafe and you wish, or feel compelled, to do something about the threat. In other words, you are already aware that a problem may exist, and you want to learn more about how to solve it.

It merits noting that the *Guidelines* documents represent a federally funded engineering innovation in earthquake safety and are designed for use in a wide variety of settings. Overall, the purpose of the *Guidelines* documents is to help you with the technical aspects of actually accomplishing seismic rehabilitation. This volume, however, explores the nontechnical factors involved in seismic rehabilitation.

Precisely because seismic rehabilitation is not a purely technical process, an often bewildering array of problems and complexities arise. Abating the risk posed by earthquake-hazardous buildings often brings into play social, economic, psychological, and various other considerations that make seismic rehabilitation very complex and, in those situations involving compliance with governmental seismic rehabilitation requirements, quite political.

SEISMIC REHABILITATION AND PUBLIC VALUES

By standard definition, politics is all about "the authoritative allocation of values" or, as one scholar put it, politics is "who gets what, when, and how." Politics, therefore, is an arena of conflict, cooperation, and compromise in which a pluralistic/democratic society, or a constituting jurisdiction, determines how and by whom a particular problem is identified, defined, addressed, and resolved — and then at what and whose cost. Given that seismic rehabilitation is really about "life safety," a central value if ever there was one, it often becomes political. Following directly from this observation, four points should be kept in mind:

First, seismic rehabilitation projects entail direct costs (e.g, engineering evaluations, the rehabilitation itself, temporary relocation), and these have to be allocated in some fashion or combination to building owners, tenants, government, and/or the public.

Second, seismic rehabilitation also entails social disruption (individual as well as neighborhood) and economic loss (foregone income). These "indirect costs," especially in urban areas, often affect the most marginal populations (the poor, minorities, the elderly) and must be borne in some way as well.

Third, it has proven inherently difficult to explain to affected populations the meaning of seismic performance levels, earthquake risk, and the effectiveness of — and trade-offs between — varying rehabilitation standards. While both direct and indirect costs are immediate, visible and have to borne by someone, the benefits of enhanced life safety are only probabilistic and rather vague (when an earthquake strikes, fewer lives will be lost); therefore, the debate often appears to suffer from misperception, misunderstanding, and shifting ground.

In fact, however, seismic rehabilitation involves values in conflict. The conflicts revolve around the trade-offs between improved life safety, a somewhat abstract concept, and very concrete costs, which are not abstract at all. Alesch and Petak (1986, pp. 66-67) capture the essence of this conflict with a quote drawn from one of the public hearings on the famous Los Angeles "Chapter 88" ordinance at which a citizen offered the following emotional observation:

> Now I've heard everything! Our brilliant City Council is going to tear down 14,000 buildings because there might be an earthquake that might knock these buildings down and the people might get hurt. So you're going to knock them down first and leave them [the people] homeless instead. That's like cutting off your arm so then you won't ever have to worry about breaking it. Are you gentlemen playing with all your marbles?

Fourth, earthquake awareness varies significantly across regions of the United States and interacts subtly with all of the above, with a normalcy bias (don't rock the boat), and with a reluctance by political leaders to being perceived as "unfair." The perception of being unfair needs explanation, however. Even if their life-safety motives are as pure as driven snow, political leaders are sensitive to this charge for it has deep roots.

The nation's founding fathers included in the *Bill of Rights* a guarantee against *ex post facto* (retroactive) legislation–that is, they expressly forbade laws that would make illegal an act that was not illegal at the time it was committed. This is a prohibition against "changing the rules after the game has been played." In the earthquake safety domain, seismic rehabilitation tends to strike this "changing the rules" nerve in our culture. It actually took a 1966 California Supreme Court decision to clear away legal obstacles for jurisdictions to require the abatement of a hazardous structure. While the particular case *(City of Bakersfield v. Milton Miller)* involved condemnation based on fire hazard, the decision provided the legal basis for subsequent retroactive earthquake programs in California. The court held:

> The fact that a building was constructed in accordance with all existing statutes does not immunize it from subsequent abatement as a public nuisance. ... In this action the City [Bakersfield] does not seek to impose punitive sanctions for the methods of construction used in 1929, but to eliminate a presently existing danger to the public. It would be an unreasonable limitation on the powers of the City to require that this danger be tolerated ad infinitum merely because the hotel did not violate the statute in effect when it was constructed 36 years ago.

The essential validity of *City of Bakersfield v. Milton Miller* was upheld in 1984 by *Barenfeld v. City of Los Angeles*, a case specifically involving earthquake-vulnerable buildings. Thus, for improved seismic safety, it seems that "changing the rules" is an inevitable byproduct of disaster learning and the impact of such learning on governmental responsibility for public safety.

Historically, earthquake disasters often have provided nasty surprises by showing entire classes of buildings to be seismically unsafe. The 1933 Long Beach earthquake demonstrated unreinforced masonry (URM) bearing wall buildings to be unsafe and the 1971 San Fernando earthquake confirmed the poor performance of these buildings and also showed that more newer "soft-stories" and "tilt-ups" were unsafe. The problem, of course, is that these types of buildings were not known to be earthquake-vulnerable or to pose life safety threats when they were originally constructed. Indeed, many buildings now deemed unsafe in an earthquake of a specified magnitude and ground motion met code requirements or at least common practice at the time of their construction. This "then/now" knowledge problem is the source of the tension between disaster learning and the political-cultural reluctance by decision-makers to be seen as changing the rules retroactively.

The most recent example of an unpleasant earthquake lesson comes from the 1994 Northridge earthquake, which revealed as vulnerable steel frame buildings, long believed to be the most earthquake-resistant type of construction. As a January 20, 1995, press release from the Structural Engineers Association of California, Applied Technology Council, and the California Universities for Research in Earthquake Engineering (SEAOC/ATC/CUREe) noted:

> The damage to . . . steel buildings has raised many serious questions for the design profession. Because many damaged structures were designed using the latest building codes and built according to modern construction practices, seismic building codes for steel construction have been essentially invalidated.

In sum, earthquakes teach, usually painfully if not tragically, but the learning generates state-of-the-art advances in earthquake engineering that, in turn, generate "guilty knowledge" about flaws in the existing building stock. The term "guilty knowledge" refers to the gap in time between the lessons disasters teach to the design professions and the corresponding policy and administrative changes. This time lag between awareness of specific risks and appropriate mitigation actions — the gap between a spot on the engineering and geotechnical learning curve and a spot on a corresponding public policy and administrative curve — has been termed "guilty knowledge." This term is a convenient way to express two different learning curves; it does not have any legal implications as used in this context (Olson and Olson, 1996, p. 30).

The increasingly sophisticated knowledge within the engineering community about weaknesses in the seismic resistance of various types of existing buildings is the moral and professional core of, and the motivator for, the *Guidelines* documents. If the engineering state of the art were static and no learning occurred, there would be no "guilty knowledge" and no need for seismic rehabilitation or, for that matter, the *Guidelines* documents and this volume. To the contrary, however, the engineering state of the art is dynamic, not static; disaster learning occurs, generating guilty knowledge: Thus, seismic rehabilitation becomes professionally important, and the *Guidelines* documents, and this volume are now necessary.

RAISING EARTHQUAKE AWARENESS

In recent years, considerable effort has been devoted to the preparation and wide dissemination by the Building Seismic Safety Council (BSSC) of provisions and technical criteria for the construction of new buildings and certain nonbuilding structures. Of particular relevance to the rehabilitation-focused *Guidelines* documents, however, was a finding from an evaluation of the dissemination process of the BSSC's new buildings resource document:

> Much of the success of BSSC's program was contingent upon first raising the target audiences' awareness of the nature of local seismic risks and of the *NEHRP Recommended Provisions* themselves. [Regarding implementation] the planning should take into account the importance of coordinating this effort with educational programs being conducted by other federal, state, regional, and local governmental agencies as well as non-profit professional and trade organizations (Nigg and Mushkatel).

Awareness was and remains the key to managing everything in the nontechnical aspects of seismic rehabilitation but especially to the approach and tactics chosen. Except for relying on normal attrition, many decisions will boil down to managing levels of anticipated conflict inherent in choosing seismic rehabilitation strategies.

ATTRITION: THE PERMANENT CONTEXT

It must be kept in mind that a regular building replacement process is ongoing in virtually every jurisdiction in the United States, a process that directly affects the earthquake-vulnerable building problem. For seismic rehabilitation, this attrition is a contextual process of building replacement that can — but not always does — make the hazardous structure problem more tractable. For attrition to have a positive effect on seismic rehabilitation, a jurisdiction must exhibit strict adherence to current codes containing seismic provisions appropriate for its seismic risk zone. The idea is to prevent the construction of new buildings of the types previously identified as earthquake-vulnerable (and of other earthquake-vulnerable classes for that matter) while the normal process of building replacement slowly reduces the number of existing earthquake-vulnerable buildings.

It might be helpful to think of earthquake-vulnerable buildings as a "stock and flow" problem. At any point in time, a jurisdiction will have a certain number of buildings that present life-safety threats in an earthquake of a specified magnitude and ground motion. That is the "stock" of the problem. Simultaneously, normal attrition processes in the community are reducing the number of vulnerable buildings, which is the "flow out" as it were. One key mitigation measure then is to prevent new, nonearthquake-resistant buildings from being constructed, which is the "flow in." In fact, in jurisdictions where an earthquake risk exists but the building codes do not have adequate seismic requirements or where the seismic requirements are not adequately enforced, the stock of vulnerable buildings may actually increase (i.e., if "flow in" exceeds "flow out," the stock of problem buildings goes up). Thus, for attrition to work positively with, not negatively against, efforts at seismic rehabilitation, a jurisdiction must keep up with the state of the art in building codes, enact them in a timely manner, and see to their careful enforcement.

Looked at from a different perspective, attrition is a race between building replacement and the recurrence interval of the appropriate "planning earthquake" for that jurisdiction. The assumption is that attrition will reduce the number of earthquake-vulnerable buildings to some acceptable minimum before the next earthquake capable of bringing them down or rendering them economically useless occurs.

For the record, assuring that attrition plays a positive role in abating the hazard posed by earthquake-vulnerable buildings is not without a level of conflict itself. Enactment and enforcement of a building code for new construction always entails debate, especially for jurisdictions that have never had a building code or seismic provisions within that code. Such conflict is usually limited to scientific and technical arguments about the existence of an earthquake hazard in that jurisdiction or, if existence of hazard is accepted, the severity of the risk. In the latter case, arguments about recurrence intervals for a specific magnitude event (the planning earthquake) predominate.

Extended attention to attrition is given here precisely because it is permanent and will play a role in every

one of the three following models of seismic rehabilitation, even in the "Mandatory Program Model." For example, in the Los Angeles program, attrition alone over the life of the program was expected to reduce the number of unreinforced masonry buildings (URMs) by 50 percent (4,000 buildings), leaving the city with only a hard core of 4,000 URMs with which to deal. As of 1991, 10 years after enacting the URM ordinance, of the URMs in Los Angeles, 53 percent had been strengthened, 17 percent had been vacated or abandoned, 16 percent had been demolished, and 14 percent were still pending action (by 1995, this may have been reduced to 5 percent according to Comerio, 1991, and personal communication, 1995).

MODELS OF ESCALATING CONFLICT

Two observations can be offered about the conflict potential inherent in the application of the *Guidelines* documents. First, the higher the earthquake awareness or "earthquake consciousness" of a region or jurisdiction, the easier it will be for proponents to explain enhanced life-safety probabilities and thereby justify and gain acceptance of seismic rehabilitation, at least as a concept. Looking back, it is not a coincidence that California has been a legislative leader in hazardous structure abatement at both the state and local levels with the most famous ordinance being "Chapter 88" of the *City of Los Angeles Building Code*.

Second, most analyses have focused on formal hazardous structure abatement programs that involve public policy directed at rehabilitating an identified set of structures. Indeed, the only book-length study is Alesch and Petak's 1986 *The Politics and Economics of Earthquake Hazard Mitigation: Unreinforced Masonry Buildings in Southern California*, which describes and analyzes the abatement efforts in (chronologically) Long Beach, Los Angeles, and Santa Ana.

In such formal or "mandatory" programs, the criteria, priorities, timetables, and costs are publicly debated — always contentiously — before the decision-makers (usually a city council) reach the final approval stage and then move into implementation. Little wonder that local governments find mandatory programs very difficult to enact and implement.

Such programs must be technically defensible, must provide for exceptions and appeals, require staff or consulting expertise, and must be perceived as not violating the "not changing the rules of the game" principle of fairness or as singling out owners and occupants of the targeted building class(es) for costly rehabilitation measures. As a result, mandatory programs tend to mobilize vocal constituencies. California examples of this type of formal program would include not only Los Angeles, Long Beach, and Santa Ana but also Santa Rosa and a few other cities.

The mandatory program idea, however, is not feasible for most jurisdictions in the United States outside California given the varying levels of seismic hazard but low levels of seismic awareness. Only in jurisdictions with relatively high levels of seismic hazard and awareness will a mandatory program proposal achieve a place on political agendas, in part because it effectively lodges at the upper end of a policy escalation ladder based on conflict potential.

There are, however, two other generic seismic rehabilitation policy options, both of which may be more realistic for much of the United States than the "Mandatory Program" model: the "Informal/Encouragement Program" model and the "Voluntary Program" model. To illustrate the level of conflict associated with the three models, see Figure 1 below which places them on a 10-point "escalation ladder."

Note, however, that this escalation ladder should not be confused with seismic rehabilitation triggers, which are discussed later and define under what conditions seismic rehabilitation requirements must be met. Rather, this ladder is a way of viewing the range of possible policy choices and sorting out their respective implications.

The escalation ladder also highlights another crucial variable — the degree of "pro-activity" exhibited by a building department. As will be explained below, in the "Voluntary Program," a building department is essentially passive. In the "Informal/Encouragement Program," a building department plays a stronger, more pro-active role, although on a selective basis. In the "Mandatory Program," however, a building

department is on the point, pushing or at least implementing surveys and program directives.

```
10   (Highest Conflict)
 9   The "Mandatory Program"
 8
 7
 6   The "Informal/Encouragement Program"
 5
 4
 3
 2
 1   (Lowest Conflict) The "Voluntary Program"
```

Figure 1 Seismic rehabilitation escalation ladder.

A slight variation of this approach reflects the complexity of the relationships between levels of government. Sometimes local officials or, more precisely, local issue advocates want the rules to be set by the state, for example, because they expect a high degree of conflict over the issue. Even if they believe seismic rehabilitation is the "right thing to do," state mandates allow local implementors to skillfully avoid conflict by explaining that they have no choice but to "carry out a state mandate."

The Voluntary Program

Not adequately appreciated is the number of buildings that have been and are being seismically rehabilitated by their owners without compulsion by local building officials. Such rehabilitation may focus on the seismic aspect alone or may feature seismic aspects as part of a larger remodeling effort. Either way, it is essentially a private or at least an owner-driven and, therefore, low-conflict process that explains its placement at conflict point "1" on the escalation ladder. Under this "Voluntary Program," owners decide, for a variety of reasons, to seismically rehabilitate their structures and approach building officials for permits and perhaps even for assistance or advice on how a building or buildings might be modified to achieve a desired level of earthquake performance. The building official then permits owners to rehabilitate the buildings on their own. Interestingly, following damaging earthquakes, voluntary rehabilitations often surge — even in jurisdictions not directly affected by the event.

The advantages of the "Voluntary Program" are considerable. Government coercion is not needed. Ordinances are not required. The media do not become involved. Motivations and decisions are largely internal. Courts and lawyers are largely avoided. Politics is seldom a factor. Community impacts are relatively minor. This approach is neither as rare nor as utopian as it might appear. Seismic rehabilitation is going on all the time in a wide variety of jurisdictions, but it occurs largely without notice except possibly within the local professional community.

Chosen from literally dozens of examples, four significant voluntary rehabilitations are described below: a public building in Utah; a private building in South Carolina; a private multibuilding complex in California; and a school rehabilitation program in Missouri, the case that best illustrates the model. Each case is different, but all share the common theme of low profile, internal decision-making and self-funding. A fifth case from Tennessee, an effort that was unsuccessful, is also described below for the sake of balance.

Voluntary seismic rehabilitation appears to occur in either of two contexts. In some cases, seismic considerations are piggybacked onto broader remodeling or rehabilitation efforts. In other cases, the seismic rehabilitation is an end in itself and is undertaken as an investment in the survival of the building against a recognized earthquake threat. The essence of the decision remains at the building level, and it is made by the owner, although mortgage and/or insurance companies also may play a role.

A special note on remodeling is in order. A remodeling effort can cut both ways for seismic resistance of a structure. While seismic strengthening obviously can be piggybacked onto remodeling, a danger lurks there as well. Unless a building official is attentive, especially in areas where earthquake awareness is low, remodeling can actually reduce the earthquake resistance of a structure depending upon how the remodeling is designed and carried out (e.g., it can weaken a load bearing or shear wall). One building official who caught such a remodeling weakening combination termed it a version of "one step forward, two steps back." The *Guidelines* documents them-

selves serve as a bulwark against such inadvertent weakening and as a resource for building officials caught in such situations.

The "Voluntary Model" contains obvious defects. First, the scope is limited only to those buildings whose owners are enlightened and/or who see long-term financial advantages in seismic rehabilitation. In other words, the rehabilitation is not systematic and depends upon financial feasibility and owner receptivity or "good citizenship." Second, the pace of seismic rehabilitation in a community is unpredictable for the same reasons. Third, the direct costs as well as the indirect costs will be passed along to the tenants, employees, and/or consumers without public discussion and, therefore, without a wide airing of alternatives and consideration of amelioration possibilities for those affected. Fourth, it is likely that the "worst" buildings, precisely because they are marginal-value properties in the first place, will not be rehabilitated by their owners, a fact that has an interesting dark side.

If we assume that seismically rehabilitated commercial and residential buildings will command higher rents, it will drive out the poorer tenants and send them toward cheaper space — very likely into those buildings whose owners have not seen fit to rehabilitate their structures. Therefore, at least in the short to middle run, it is possible that voluntary seismic rehabilitation may actually increase the population concentration at risk in other (unrehabilitated) buildings.

In addition, seismic rehabilitation and its costs are only inputs into a larger decision. While the *Guidelines* may offer seismic rehabilitation goals, techniques and cost estimates, other factors may prove decisive, especially if the total rehabilitation project costs outweigh new construction costs.

In total, the case studies illustrate that while the *Guidelines* documents will be extremely useful, many other factors often will be present. As appealing as voluntary approaches are, there are some serious risk perception and economic obstacles to their more widespread use. Among them are individuals' estimation of the probability of an earthquake damaging their structure being sufficiently low that the investment in rehabilitation will not be justified; the tendency to assign very high discount rates to such decisions, which results in giving future benefits very little weight compared to spending money for protective measures; and judgments that current prices for seismic rehabilitation measures simply are too high, to even focus on the potential value of reducing future losses. Such determinations are likely based on arguments having little to do with expected benefit/cost comparisons.

Case 1: The 1894 Salt Lake City/County Administration Building

Salt Lake City, like all major population centers in Utah, sits astride the Wasatch Fault at the base of the Wasatch Mountains. The fault is considered historically active but so far has not done major damage to the urban areas of Provo, Salt Lake City, or Ogden. The U.S. Geological Survey and the Utah Geological Survey consider the earthquake threat to be serious.

In the late 1980s, Salt Lake City faced the problem of what to do about its earthquake-vulnerable but historically and architecturally valuable Administration Building. The decision was made to seismically rehabilitate it using a "base isolation" method. The rehabilitation was undertaken voluntarily and paid for by the city to protect a major asset and to serve as an example of government leadership and responsibility in seismic safety.

Case 2: The North Charleston Hotel

A major hotel chain faced an interesting problem after constructing a new hotel in the city of North Charleston, South Carolina. At the time of construction, North Charleston had no specific earthquake-resistance requirements in its building code, in large measure because the state did not have (and as of May 1996 still did not have) a building code.

After construction of the hotel, however, a national insurance company would not accept the mortgage because it had evaluated regional seismic risk (hardly a secret given the 1886 event) and noted the lack of an appropriate seismic component in the original design of the building. The insurance company then commissioned a San Francisco engineering firm to recommend a rehabilitation plan that would meet the company's earthquake performance

requirements for the region. Subsequently, an external steel frame that tied back into the original concrete frame was added to the hotel. In short, the investment — or more precisely, the collateral — was protected.

All of the key decisions were made in the private sector. This case provides an important perspective on how the insurance industry, banks, and other financial institutions and the building and real estate communities could work together to foster seismic rehabilitation with or without governmental participation.

Case 3: The PG&E Buildings, San Francisco

The Pacific Gas and Electric Company (PG&E) is headquartered in San Francisco and has a long and colorful history in "The City." At an approximate total cost of $150 million, PG&E chose to seismically rehabilitate a complex of four of its older office buildings partly using the benefits of the Preservation Tax Incentives for Historic Buildings. The rehabilitation was reviewed by the California State Office of Historic Preservation and the National Park Service and certified as meeting the Secretary of the Interior's Standards for Rehabilitation, thus earning a 20 percent investment tax credit (approximately $30 million).

The motives were four: to remain in the city, to save landmark structures facing the famous Market Street, to protect PG&E employees, and to set an example in the community of a voluntary business commitment to earthquake safety in general and to seismic rehabilitation specifically. The details of this case are especially interesting. According to representatives of PG&E's structural engineering consultants (Jokerst and Elsesser, EERI, 1995):

> *The complex of four pacific Gas and Electric Co. Office Buildings in downtown San Francisco built from 1921 to 1949 represent a variety of multi-story construction ranging from 9 stories to 18 stories and encompass over 500,000 square feet of floor area. These buildings are part of an essential complex for the public utility which provides natural gas and electricity to Northern California. After the 1989 Loma Prieta earthquake, which caused limited damage to the buildings, PG&E determined that a seismic upgrade of these four old*

steel frame buildings was justified to meet the cor- porate goal of being operational after a strong earthquake.

Ten seismic strengthening options were studied for the two primary 18-story L-shaped buildings form- ing the center of the complex. Each alternate was evaluated to determine its impact on (1) interior space planning, (2) historic features, (3) dynamic response, (4) capacity of existing foundation, (5) existing frame capacity to support the increased seismic loads, (6) pounding between the adjacent structures, and (7) lateral drifts.

The PG&E complex demonstrates a performance- based approach to design which goes beyond the simple code-based life safety methods. This project addresses the desire by Pacific Gas and Electric Company for a facility which will serve the public after the next damaging earthquake.

Case 4: A Missouri School District

A special version of the "Voluntary Program" is ex- emplified by officials of the School District of Clay- ton, Missouri. Part of the greater St. Louis area, the District needed a voter-approved $6.6 million bond issue to finance new or replacement construction and a range of school improvements. These officials recognized the earthquake threat in the New Madrid area but understood equally well that the public threat perception was low. By "packaging" seismic considerations as one of the five "compelling and immediate needs" inside an overall bond argument, however, the Clayton School District won the bond election and was able to carry out nearly $3 million of seismic rehabilitation projects "by strengthening portions of existing schools."

Case 5: Memphis, Tennessee

The first four cases and the description of the Volun- tary Model tend to bias perception in that only "suc- cess" stories are told. As a partial balance to this somewhat excessive optimism, consider the story of a major automobile parts and accessories chain with headquarters in Memphis that evaluated its present location in a structure designed originally as a de- partment store. Seismic performance was explicitly included in the overall rehabilitation evaluation;

however, in the end, the company chose to construct a new building with appropriate seismic design in the downtown area because all things considered, constructing a new building was actually less costly than rehabilitating the old one. If, as in this case, the total project cost outweighs that of constructing a new building, seismic rehabilitation most likely will not be occur.

The Informal/Encouragement Program

Like the voluntary approach, the "Informal/En- couragement Program" is more common than is of- ten appreciated. Although not commonly acknow- ledged, building officials often try to reach agree- ment with owners involved in building rehabilitation. Such negotiations can be based on authority granted by local ordinance or can be conducted as part of a building official's administrative responsibilities. This is because each building "has its own story."

A former midwestern city building official com- mented that "in contrast to new construction, negotia- tion is a way of life in dealing with existing build- ings, and the architect/engineer/owner could walk away from negotiation or use a board of appeals pro- cess." This approach involves a building official ne- gotiating seismic considerations into an owner's re- quest for permits to remodel an existing structure. In this case, an owner requests permits to do various kinds of work on a structure, and a local building official says in effect, "Okay, but you also have to include some seismic rehabilitation measures as well." Four example cases are presented below.

Case 6: Provo, Utah

The city of Provo, which like all other cities in Utah sits along the Wasatch Fault, achieves seismic reha- bilitation of existing buildings by negotiation with building owners. No mandatory requirements exist to require the seismic rehabilitation of URM build- ings. The building department applies its negotiated informal approach only when a significant improve- ment or change occurs to one of these buildings, most of which are located in the older central busi- ness district and date from the late 1800s.

The standard for URM building strengthening in such cases is the current Uniform Code of Building Conservation (UCBC), Appendix Chapter 1. Example alterations that affect structural elements or increase loads include adding to a mezzanine or changing uses that would increase floor live loads. When an agreement is reached between the building official and the owner on the scope of the seismic rehabilitation effort, the official issues the permit.

In recent years, however, none of the subject buildings has had any alterations proposed that would trigger discussions about seismic rehabilitation. It is possible that once an owner becomes aware that the city might require seismic strengthening, the scope of the proposed project is changed to avoid such work or, in some cases, the project is canceled. In some cases, it may be that the requirements for seismic rehabilitation, albeit negotiated informally, are sufficient to deter some significant property improvements in the area.

It is interesting to note that in 1995 Provo's building department proposed a mandatory parapet bracing requirement. Principally because of cost concerns, the proposal never got far enough along in the policy process to reach the city council. Interestingly, the council has rather deftly stayed on the sidelines in discussions related to building codes. It generally defines code issues as "technical" rather than more broadly political, thus containing the debates within a relatively narrow circle of building officials and other stakeholders and interested individuals.

Nevertheless, some progress is occurring. In addition to URM buildings, when improvements or additions are made to wood frame buildings, the city looks for evidence that the wall sill plates are anchored to the foundation or slab. If these connections do not exist or are less than the code required minimum, the city requires new anchors (sill bolts) to be installed as a condition of the permit.

Case 7: Seattle, Washington

When a building undergoes substantial remodeling in Seattle, seismic rehabilitation is mandated. The extent of the improvement in its seismic performance can be negotiated, however, under the following 1995 revision to the Seattle Building Code:

***3403.3 Impracticality.** In cases where total compliance with all the requirements of this code is impractical, the applicant may arrange a pre-design conference with the design team and the building official. The applicant shall identify design solutions and modifications that conform to Section 104.14. The building official may waive specific requirements in this code which he/she has determined to be impractical.*

Section 104.14 states that an "alternate" may be approved by the building official if he/she finds that it "complies with the provisions of this code and that the alternative, when considered with other safety features of the building or other relevant circumstances, will provide at least an equivalent level of strength, effectiveness, fire resistance, durability, safety and sanitation."

Case 8: Palo Alto, California

Home to Stanford University and many high technology companies, the 55,000-person city of Palo Alto recognized its earthquake-vulnerable buildings problem and has taken a unique approach to seismically rehabilitating these buildings. After a lengthy exploration and negotiation process, the city adopted a "Seismic Hazard Identification Program." It does not fall neatly into any program category, but mostly resembles the "Informal/Encouragement Program" because some of the program's elements are mandatory while others are voluntary and incentive oriented.

Palo Alto's efforts to deal with its vulnerable buildings date from the mid-1970s, but it was the 1983 Coalinga earthquake that led to the creation of a Seismic Hazard Committee "representing a diversity of interests" (stakeholders), which ultimately agreed upon the scope of the existing program. The key elements of Palo Alto's program are:

- *It imposes rehabilitation requirements on 99 structures in three categories (all URM buildings, all pre-1935 non-URM buildings with 100 or more occupants, and all buildings with 300 or more occupants constructed between January 1, 1935, and August 1976).*

- *Once notified by the city, the buildings' owners are required to contract with a structural engineer. Given a specified time period in which to* conduct a study and file a report with the city, the owners' engineers have to evaluate the earthquake vulnerability of the building and to identify what should be done structurally so that the building will meet the seismic provisions of the 1973 Uniform Building Code (UBC). The reports are reviewed by consulting engineers to ensure they comply with the ordinance.

- Each building owner must notify the occupants in writing that an engineering report has been completed and that the report is available for review in the city's Building Inspection Division.

- Within one year after filing the engineering report, each building owner also must submit a letter indicating his/her intentions regarding correction of seismic deficiencies. Failure to comply could result in injunctive relief, criminal prosecution, or both.

The underlying policy philosophy was that "while no mandatory retrofitting (rehabilitation) requirement was imposed . . . the reporting requirements would create sufficient concerns about liability and about the decline in the market value of earthquake-deficient structures, that seismic improvements would occur voluntarily" (Beatley Berke, pp. 63-64).

Some clues are available about the implementation of the program:

- A downtown density and parking incentive are provided for seismically rehabilitated buildings. Bonuses are given for the buildings in the three categories that exempt them from providing on-site parking as a condition of rehabilitation.

- Compliance with the reporting requirements has been good — virtually 100 percent.

- The reports and public disclosure requirements — reinforced by California's real estate disclosure laws on property sales and purchases — act as strong incentives and a number of seismic upgrades have been completed.

- Some tenants in leased buildings have helped finance the seismic upgrades through lump-sum payments or higher lease costs, and others have agreed to vacate before and return to the building after the seismic rehabilitation project is com-

pleted. This protects the owners' abilities to service their debts.

- Some innovative developers have found ways to capitalize on the seismic rehabilitation program by publicizing the work done, taking advantage of the greater square foot allowances provided under the parking incentive measure, and even trying to obtain the bonus for buildings not in the three covered categories.

- Early fears that owners would be unable to continue to insure their governed properties for liability are not being borne out. Increases in rates, however, are a possibility.

- The private owners are carrying the direct costs of the program's reports and seismic rehabilitation improvements.

An interesting sidebar to Palo Alto's program that may have reinforced private owners' willingness to accept the ordinance was that the city voluntarily seismically rehabilitated its Civic Center building. This structure was constructed between 1968 and 1970 and is an eight-story tower supported by a three-story below-grade parking structure. The project was financed by "Certificates of Participation," and the work was done in slightly more than two years "while the building was occupied and in full operation" (Sharpe p. 1).

Case 9: San Leandro, California

The 15 square mile Alameda County city of San Leandro borders Oakland on the north and is a mixed residential, commercial, and industrial area of about 70,000 mostly middle-income residents. The eastern part of San Leandro spans the active Hayward Fault. San Leandro has dozens of URM buildings, thousands of older wood-frame dwellings, modern apartment structures, and tilt-up light industrial buildings along the San Francisco Bay's shoreline, all of which are earthquake-vulnerable.

The city's earthquake safety efforts — triggered by the recommendations of a citizen task force — demonstrate an interesting voluntary government-citizen partnership. Known as the "1993 Seismic Retrofit Financing Project," the city council approved raising $12,780,000 through "Certificates of Participation" to seismically strengthen several municipal buildings. The buildings included rehabilitating the 1965 City Hall, the 1970 South Office Building, and the 1968 Public Safety Building, which houses San Leandro's fire and police departments and their communications and dispatching centers.

In addition, the city has supported seismic rehabilitation by its residents. Part of an annual $300,000 earthquake preparedness appropriation (which includes federal mitigation grant funds) assists residents with the strengthening of their homes. Detailed easy-to-understand instructions are provided to owners by the building department; classes are provided by qualified engineers; tools are loaned to property owners; the work is inspected at no charge; and the property owner receives certification that the building has been strengthened to the city's standards.

In general, the "Informal/Encouragement Program" would have to be marked as medium-conflict ("5" or "6" on the escalation ladder) because, no matter how informally the seismic requirements are leveraged in, it is a form of government mandate to have seismic rehabilitation included as a "must be" part of an overall permit process. Under this model, a building department is obviously proactive, not passive, but in a selective manner.

In practice, when a jurisdiction employs this approach, building owners tend to complain that the city building department is being "unreasonable." While probably rare, attempts at political end-runs to a city council, mayor, or city manager could be made to test the resolve of the building department — and its political support. Seattle's experience is that almost no appeals have gone to its mayor or council. This is because its seismic rehabilitation triggers (when is rehabilitation required) are specified in ordinances even though the extent of the rehabilitation work involved is negotiated. In general, it is both clear and prudent that building departments have some reference standard, such as the UCBC or formally adopted ordinances, to avoid the potential nightmare of inconsistent and capricious requirements being imposed. At the same time, however, formal rehabilitation ordinances are not required, neither the media nor the courts tend to be involved, and the political conflict generated remains con-

tained within a fairly small circle of officials, owners, and engineers. In other words, seismic rehabilitation does not become an explosive public issue, which is often the case with the upper end inhabitant of the escalation ladder, the "Mandatory Program Model." Finally, owners may abandon their projects or redefine them to avoid triggering even informal requirements. A common way of doing this is to perform a series of smaller projects that do not trigger seismic rehabilitation but that collectively result in a major alteration.

The Mandatory Program

As indicated above, the "Mandatory Program" is definitely high-conflict and rates a kind of general "9" on the ladder, but it could range anywhere from "8" to "10." For example, if the number of buildings targeted in a jurisdiction is relatively small and if the required rehabilitation is at least partially subsidized (e.g., through a redevelopment project), the score could be an "8." On the other hand, if, as in the famous Los Angeles case, thousands of buildings are involved and no external financing is offered, the program can — and did — reach a "10" on the conflict ladder. In essence, mandatory seismic rehabilitation programs are full blown public policy. As such, formal ordinances stipulate priorities, criteria, processes, choices, rules, coercive measures, timetables, and even appeal processes. Moreover, given the very public nature of the decision-making, the process is long, arduous, and very political.

Not only does a "Mandatory Program" debate entail extended technical arguments, it also gives at least equal time to the direct cost question (how much for what level of safety), the cost incidence question (who pays initially but who pays in the end), and the indirect cost considerations (differential impacts on marginal populations, personal disruption, neighborhood effects). Battles also are joined on scope (what buildings), priorities (which buildings first and why), and pace (how fast). Most important, a mandatory program stimulates the creation of what once were called "interest groups" but now are more accurately referred to as "advocacy coalitions" or "stakeholders," each having its agenda or special focus. As a result, the media and the courts become involved, often sooner rather than later.

In the "Mandatory Program," seismic rehabilitation is imposed coercively on building owners by government, and most of the politics revolves around attempts by the owners to minimize the scope and requirements of seismic rehabilitation and, therefore, the costs. Owners then attempt to externalize (shift to others) those costs to the greatest degree possible. The decision arena is usually a city council, and mandatory programs tend to involve not only the elected officials but also numerous individuals and groups including building owners, tenants, building safety officials, professional engineers, historic building advocates, neighborhood organizations, and even representatives of other levels of government. The "pro" and "con" sides (advocacy coalitions) become very complex. In a discussion separate from his book with Alesch, Petak offers a summary of the kinds of actors involved in the development and passage of the hazardous structure abatement ordinances in Long Beach, Los Angeles, and Santa Ana (see Figure 2).

In addition to its own intrinsic conflicts, any proposal for a formal seismic rehabilitation program must face "extrinsic" challenges. That is, aside from all the internal debates, seismic rehabilitation using the mandatory approach must compete with other community priorities for scarce public funds, even if only for enforcement costs. These costs should not be underestimated in that they often entail new responsibilities for a building and safety department and very likely for the city attorney's office and planning and housing departments in larger cities.

Case 10: Long Beach — It Led The Way

As a result of the major earthquake of 1933 which bears its name, the city of Long Beach amended its building code in January 1934 to effectively prohibit any future construction of unreinforced masonry buildings, hundreds of which suffered serious damage in the earthquake. This policy was extended statewide by the Riley Act, which was passed in 1934 by California's Legislature.

Nothing was done about existing URM buildings in Long Beach until 1959 when a true hero of local efforts at seismic safety, building official Ed O'Connor, took advantage of a theater relicensing

controversy to push through an ordinance giving the building department the authority to "determine by inspection if an existing building is substandard or constitutes a nuisance" and, if so, to order the building repaired, vacated, or demolished. Once a 1966 California Supreme Court decision (City of Bakersfield v. Milton Miller) cleared the way by determining that it was unreasonable to hold cities hostage to old buildings given "the fact that a building was constructed in accordance with existing statutes [at the time of its construction] does not immunize it from subsequent abatement as a public nuisance," O'Connor attempted to implement the original Long Beach ordinance. A political uproar ensued, and while the URM problem was "studied" at length, effective implementation of the ordinance was tabled, but it at least had gone through the formal hearings process.

Major damage to URMs in the 1971 San Fernando earthquake rekindled Long Beach's interest in its URM problem and on June 29, 1971, the Long Beach City Council passed a specific ordinance to abate the hazard posed by earthquake-vulnerable structures in the city. Implementation was slowed by complexities in the ordinance such as the assignment of "hazard points," which was confusing to the owners. O'Connor argued that it was very difficult to enforce an ordinance with multiple choices. In 1976, an amendment established a more formal but simpler program with criteria for a building-by-building "hazard index" and with timetables for surveys, notifications, evaluations, and abatement. Eventually, almost 900 pre-1934 masonry, concrete, or steel buildings were either seismically rehabilitated or demolished. Thus, while Los Angeles may be more famous, its neighbor, the City of Long Beach, led the way.

Case 11: Los Angeles — The Most Famous

Although "guilty knowledge" about the earthquake vulnerability of URM buildings had existed for several decades (at least since the 1933 Long Beach event) and although the city of Long Beach itself had been working on the earthquake-vulnerable building problem since 1959, it took the devastatingly concentrated life loss of the 1971 San Fernando event (47 of the 54 fatalities took place in portions of the

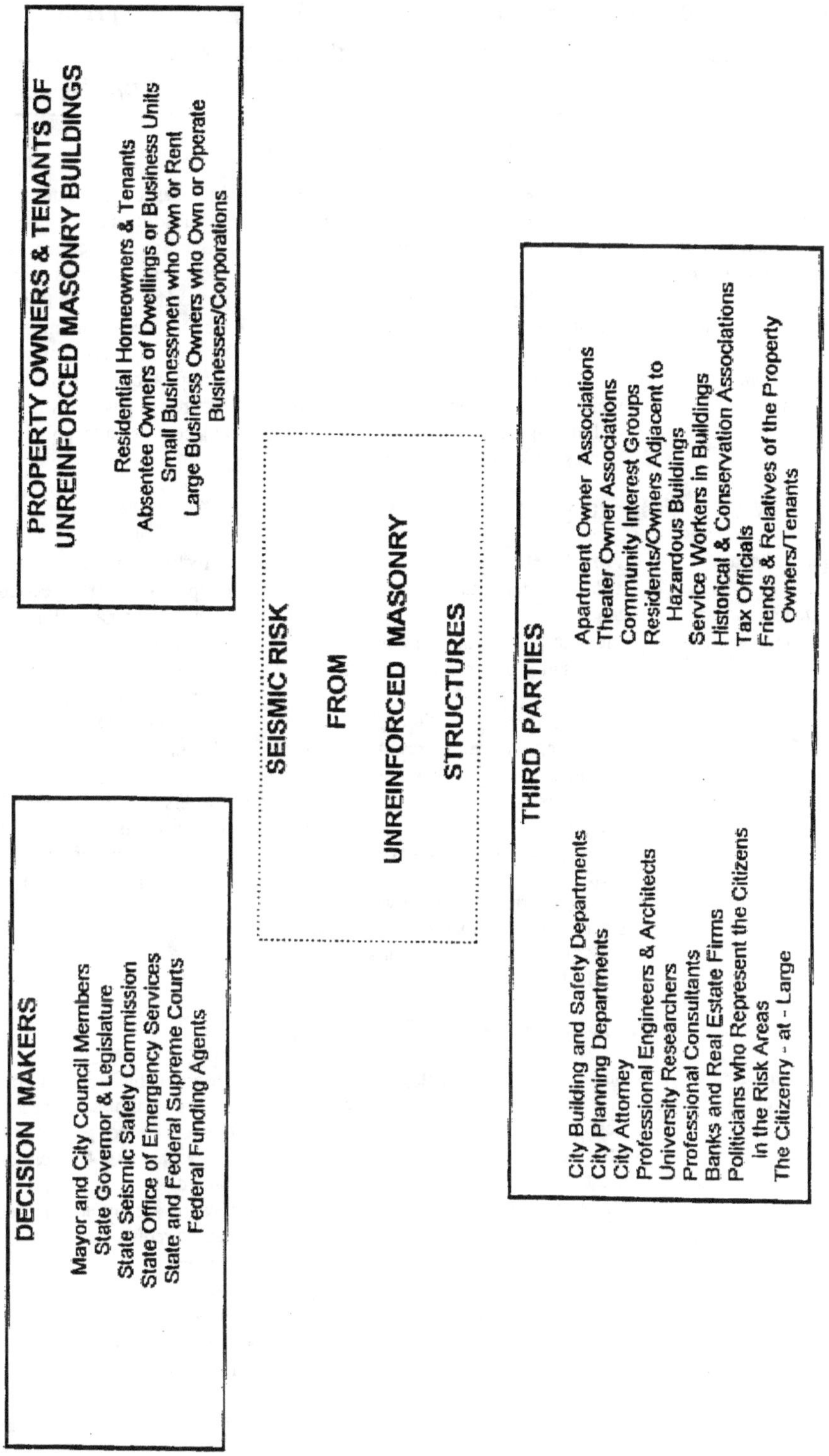

PROPERTY OWNERS & TENANTS OF UNREINFORCED MASONRY BUILDINGS

Residential Homeowners & Tenants
Absentee Owners of Dwellings or Business Units
Small Businessmen who Own or Rent
Large Business Owners who Own or Operate
Businesses/Corporations

SEISMIC RISK

FROM

UNREINFORCED MASONRY

STRUCTURES

DECISION MAKERS

Mayor and City Council Members
State Governor & Legislature
State Seismic Safety Commission
State Office of Emergency Services
State and Federal Supreme Courts
Federal Funding Agents

THIRD PARTIES

Apartment Owner Associations
Theater Owner Associations
Community Interest Groups
Residents/Owners Adjacent to
Hazardous Buildings
Service Workers in Buildings
Historical & Conservation Associations
Tax Officials
Friends & Relatives of the Property
Owners/Tenants

City Building and Safety Departments
City Planning Departments
City Attorney
Professional Engineers & Architects
University Researchers
Professional Consultants
Banks and Real Estate Firms
Politicians who Represent the Citizens
in the Risk Areas
The Citizenry - at - Large

Figure 2 A sampling of parties concerned with city seismic regulation development (from W. J. Petak).

FIGURE 3
Advantages and Disadvantages of Major Types of Mitigation Programs
for Unreinforced Masonry Buildings

Program Description	*Advantages*	*Disadvantages*
Mandatory Strengthening Programs		
• Requires owners to reduce earthquake hazards within established time frames • Timeframes for compliance start when an order is issued by the Building Department • Establishes seismic retrofit technical standards • Sets a goal of hazard reduction, not total elimination of the hazards	• Local governments can effectively enforce the program and reduce hazards • Building departments can monitor and report progress • Building departments can control compliance rates by slowing down or speeding up the issuance of orders to building owners • Compliance rates vary with the number of building occupants, with longer time frames for smaller buildings	• Imposes arbitrary and at times inflexible deadlines on building owners • Compliance schedules do not necessarily reflect the limits of the local design and construction industry resources • Can impose economic hardships on owners and occupants • Compliance schedules do not consider hazards to passersby or hazards from adjacent or unoccupied buildings.
Voluntary Strengthening Programs		
• Requires owners to prepare hazard evaluation reports • Requires owners to write letters that indicate their intentions to reduce hazards • Reports and letters are made available to the public • Establishes seismic retrofit technical standards • Owners set their own time frames for compliance with standards	• Provides effective disclosure of hazards to owners and in some cases to tenants • Flexible time frames for compliance can result in fewer economic difficulties • Rates of hazard reduction can vary depending on owner's resources and demands on the design and construction industry • Provides an effective management and monitoring system to local governments • Local governments can always reconsider the program's progress and impose mandatory requirements if it is ineffective.	• Effective in reducing hazards only if coupled with strong economic environments, and financial, planning, and zoning incentives • Not effective with owners who choose not to cooperate, and thus can be unfair to cooperative owners • May prolong overall hazard reduction efforts and earthquake risk exposure • Owners must pay higher fees to design professionals • Does not consider hazards for occupants and passersby or from adjacent buildings
Notification-Only Programs		
• Owners are notified by letter that their buildings are potentially hazardous	• Some local governments state that it meets the minimum intent of the URM Law • Minimal initial cost to local governments • No direct cost to owners who choose to ignore hazards • Can be effective if owners are few and cooperative and if governments adopt seismic retrofit standards	• Programs have been ineffective in reducing earthquake hazards • Owners are not protected from future code changes if they choose to reduce hazards • Owners are not encouraged to consider hazard reduction • Owners are not informed of specific hazards and are likely to react with disbelief • Local government can't easily monitor hazard reduction progress • Imposes demands on local governments to deal with unhappy owners • Seismic retrofit standards are typically not adopted

Veterans Administration hospital built in 1925) to force open a political window of opportunity for seismic rehabilitation in Los Angeles in February 1973. The scale was daunting — the estimate was that the city had 14,000 earthquake-vulnerable buildings. A key actor once described the problem as: "How do you eat an elephant? Well, one bite at a time." Befitting the "Mandatory Program" model, debate over various versions of the hazardous structure abatement ordinance became very contentious very rapidly with building owners mounting strong attacks against each draft. Alesch and Petak (1986, p. 62) quote a leader of a group of apartment owners who captured almost all (he missed historic preservation) of the principal objections in a single diatribe:

> *The proposed ordinance is a direct attack on the poor . . . on senior citizens . . . on every tenant in the city . . . makes it impossible for the owners of and investors in the older buildings to comply with it . . . would put tremendous upward pressure on rents in the city . . . create unimaginable voter unrest . . . .*

After three years of conflict, the Los Angeles city Council sent a draft ordinance back to committee for further study in December 1976.

Advocates for an ordinance regrouped and found a city councilman (from the area most damaged by the 1971 San Fernando event) who took the public and political lead and guided the next version of the ordinance, which would become Division 88 of the Building and Safety Code, through a continuously acrimonious process to final passage on January 7, 1981. Almost eight years elapsed between placement of the earthquake-vulnerable buildings problem on the political agenda in Los Angeles and final passage of the ordinance.

Case 12: State of California Senate Bill 547 (and Senate Bill 445)

In June 1986, the Governor of California signed into law Senate Bill (SB) 547. This law require cities and counties in Seismic Zone 4 (which included approximately 80 percent of California's population) to inventory their URM buildings and, by January 1, 1990, to establish programs to mitigate the hazards they posed. For many jurisdictions, the results of the inventories were an unpleasant surprise and constituted the first solid information they had on the extent of their URM building problem. Because of SB 547, many jurisdictions suddenly had "guilty knowledge" about earthquake-vulnerable URM structures in their building stocks.

While SB 547 did not specify precisely what mitigation programs had to be put in place by the local jurisdictions, in 1991 the California Seismic Safety Commission (CSSC) identified the four types that had evolved: mandatory strengthening, voluntary strengthening, notification only, and "others." Not surprisingly, the CSSC preferred the mandatory approach, saw advantages in the voluntary program, but had serious reservations about the "notification only" program. The "others" were too varied to cover easily. The CSSC then outlined the advantages and disadvantages as they saw them of the three major types of URM mitigation programs (Figure 3).

Although enacted seven years earlier than SB 547, another law, SB 445, should be mentioned. SB 445 allowed local governments in California to adopt standards for seismic rehabilitation of URM buildings that were lower than the standards for new construction. SB 445 had a dual effect: It reduced estimates of the rehabilitation costs for URM buildings (because repair could be to a lower standard) but, more important, it removed local government concern about legal liability for having different standards for rehabilitation of existing buildings and new construction.

Case 13: Seattle—Changing Focus and Local Policy

The city of Seattle's experience illustrates how the failure of a mandatory retrofit ordinance led to the current negotiated methodology. In essence and for a variety of reasons, Seattle's policy moved from a focus on one area (the historic "Pioneer Square") to all business districts where parapets are common hazards and finally to a triggered mandatory requirement that applies to all existing buildings but that allows for negotiation of the level of structural improvement on a case-by-case basis.

"Pioneer Square" is a 15-square-block area adjacent to Seattle's central business district. Its buildings (largely URM) were constructed at the turn of the century. It provides an example of the difficult-to-implement mandatory rehabilitation policy for a specific district. In 1973, ordinances were passed that applied solely to the Pioneer Square Historic District. They specified minimum maintenance requirements and also required rehabilitation of the URM buildings (to ensure that all structural members could "carry imposed loads with safety" and prevent any portion of the exterior from falling in an earthquake). "Substandard historic building" notices were sent out, and by May 1977 only 18 out of 143 buildings had been partially rehabilitated buildings rehabilitation. Further achieving the necessary increased rents to pay for the improvements was often unrealistic. Lengthy hearings were required before the building department could take enforcement action and, as a result, the rehabilitation requirements were repealed and strengthening requirements were triggered only if a building was to be substantially remodeled.

In November 1975, a large section of terra cotta cornice tile fell from a multistory building onto a sidewalk near the downtown retail core. This event initiated a formal inspection and notification program for Seattle's central business district, in particular the entire downtown core. This was followed by adding new language to the 1977 Seattle Building Code that specifically required abatement of "unsafe building appendages" like URM parapets. An inspector/engineer was assigned to try to identify all such hazardous parapets (many of which were in Pioneer Square). Most of the hazardous parapets in the downtown area (including Pioneer Square) had their parapets braced. This ordinance is still used on URM buildings outside of the downtown area.

Thus, the mandatory requirement for the "global" (although "partial" in current engineering terms) rehabilitation of URM buildings failed, but a very modest mandatory requirement for strengthening one of the URM buildings' most widely recognized hazards (parapets) has been very successful.

A useful and successful example of seismic rehabilitation policies is Seattle's current one that applies to all existing buildings. When an existing building undergoes a "substantial remodel" (remodeling that extends its "useful physical and economic life"), its seismic risks must be mitigated. This trigger (and there are a couple of less frequent ones) is codified, not negotiated. There is usually a pre-design meeting with the owner, the engineer, and specialized building department staff. At this meeting, the level of structural improvements is negotiated, the goal being to ensure that the degree of improvement is "commensurate with the size and scope of the proposed project." Thus, the rehabilitation is mandatory (as triggered by a proposed remodel), but the level of structural improvement varies from case to case. This has been very successful for many years, and a wide variety of office, retail, light manufacturing, and residential (including low income) buildings have been rehabilitated.

Case 14: San Francisco's "Bolts-Plus" Partial Rehabilitation for Unreinforced Masonry Buildings

Passage of California's URM law in 1986 (Chapter 12.2, Section 8875 et. seq., "Building Earthquake Safety" of the Health and Safety Code) accelerated local government consideration of the URM problem. In San Francisco, this process ultimately resulted in the passage of San Francisco's Ordinance 225-92, on July 13, 1992, "relating to earthquake hazard reduction in unreinforced masonry bearing wall buildings." With the avowed primary social purpose of preserving low-cost housing, the ordinance has lower safety standards than the state-adopted model code (discussed below) when applied to normally configured residential occupancy buildings. Ordinance 225-92 allows residential and certain commercial use unreinforced masonry buildings (UMB in San Francisco terminology) to be rehabilitated using a "bolts-plus" solution ("the installation of shear and tension anchors at the roof and floors and, when required, the bracing of the UMB walls upon evaluation of the height-to-thickness ratio of these walls, Section 1603B1.1). This method cannot be used for buildings housing assembly, educational, or hazardous occupancies as defined in the building code.

The process of establishing the technical basis for Ordinance 225-92 is worth some discussion. As noted above, the state's URM law required local governments in Seismic Hazard Zone 4 to identify (inventory) the quantity of URM buildings in their jurisdictions, to prepare a plan to mitigate the hazards, and to file a report on their actions with the California Seismic Safety Commission (CSSC). San Francisco identified 1,967 masonry bearing wall buildings. (Approximately another 120 nonbearing wall URM buildings also have been identified by San Francisco, but they are outside the scope of its retrofit ordinance.)

In late 1988, San Francisco officials asked the Structural Engineers of Northern California (SEAoNC) to develop guidelines that could be used to prepare a city ordinance. SEAoNC appointed an ad hoc committee for this purpose. About the same time, the CSSC asked the counterpart statewide organization, the Structural Engineers Association of California (SEAoC), and the California Building Officials (CALBO) to help the Commission update its model ordinance focusing on bearing wall URM buildings. First published in 1985, the original basis of the model ordinance was Los Angeles' Building Code Division 88. The model was revised in 1990, 1991, and 1995. It is known now as the "1995 Recommended Model Ordinance for the Seismic Retrofit of Hazardous Unreinforced Masonry Bearing Wall Buildings."

Part of SEAoC's and CALBO's response to the CSSC was to convert the technical provisions of the model ordinance into a format acceptable to the International Conference of Building Officials (ICBO) for use in all seismic zones. The technical provisions of the revised model ordinance became Appendix Chapter 1 to the 1991 edition of the Uniform Code for Building Conservation (UCBC), a companion document to the Uniform Building Code (UBC). The administrative provisions of the model ordinance are not included in the UCBC. In 1991, the State of California adopted the UCBC's Appendix Chapter 1 as a model code.

The issue was referred to an advisory committee, the Seismic Investigation and Hazards Survey Advisory Committee (SIHSAC), which was established about 1980. In addition to engineers and architects, it was composed of contractors, real estate and lending interests, and others. While the SIHSAC generally agreed that the UCBC was an appropriate ap-

proach, strong opposition came from UMB property owners, especially those in lower income, rental rate, and property value areas of San Francisco. This led to two important studies — an environmental (and economic) impact report and benefit-cost analyses of UMB rehabilitation alternatives. These reports were used by a largely nontechnical task force (discussed below) to fashion a politically acceptable compromise. The SEAoNC's ad hoc committee recommended that San Francisco adopt California's new model code.

The opposition to the UCBC approach led the Board of Supervisors and the Mayor of San Francisco to form a two-part task force to review the SIHSAC's recommendations. The task force, composed of representatives of several city departments and other organizations (assisted by a 40+ member Community Advisory Committee) recommended allowing the "bolts-plus" approach because, at least for normally configured buildings, this would prevent 80 percent of the URM building earthquake life-safety problem (out-of-plane failure of the bearing walls). Ultimately, this became the political selling point of Ordinance 225-92. Ironically, however, some engineers believe that only a small percentage of all the inventoried unreinforced masonry buildings are actually eligible for "bolts-plus" rehabilitation.

The Loma Prieta earthquake on October 17, 1989, accelerated the process of enacting the UCBC as a state model code (not necessarily a minimum) for rehabilitating URM buildings (Chapter 173 of the 1991 Statutes, which amended several individual state code sections). Meanwhile, the SEAoNC used Loma Prieta's "window of opportunity" to get some significant limits on the use of "bolts-plus" inserted into San Francisco's pending Ordinance 225-92. For example, the "bolts-plus" rehabilitation method cannot be used on a URM building unless it has a regular configuration, has qualifying cross walls, and has a specified minimum area of solid URM wall.

One participant in this process noted that Ordinance 225-92 was "totally driven by socioeconomic issues." Ordinance 225-92 states: "UMBs are vital to San Francisco's economy. They provide low-cost housing, job sites, and irreplaceable historic and architectural resources. Yet, in an earthquake, they pose *a great danger to passersby and occupants." UMB structures also continue to expose low-cost housing to a sudden and permanent loss of habitability after moderate to major ground shaking even though their risk to life is reduced.*

Notices regarding compliance and "inventory forms" were sent to the owners of the governed buildings. Dates for subsequent compliance with the ordinance's rehabilitation provisions were staggered depending on the perceived relative hazards of a building's location, size, and occupancy. Compliance dates ranged from 3.5 to 13 years. If owners do not comply within the specified time period, the city's final recourse is to condemn the building so it cannot be used.

With strong support from the Board of Supervisors, in 1992 San Francisco voters overwhelmingly approved a General Obligation Bond issue of $350 million "to help owners of seismically unstable buildings finance retrofitting. . . . " While required rehabilitation is under way, as of October 1996 little of the money has been committed because: (1) commercial loans or private financing is available in a healthier economy, (2) administrative requirements are too burdensome or add to the potential costs, (3) some owners are postponing work until "the last possible minute," and (4) financing of some projects is complicated because of the need to integrate the seismic rehabilitation financing with other low-income housing financial and regulatory measures.

REHABILITATION POLICY CHOICES: OTHER CASES

Central to the overall purpose of the *Guidelines* documents is the provision of a framework to help users understand and then select desired levels of seismic performance of buildings. As the user will note in Volume 1 of the *Guidelines*, a user must select, for every structure which is a candidate for rehabilitation, a specified level of desired performance. Historically, these types of decisions have been based on preparatory technical studies or, more subjectively, on the feasibility of the rehabilitation. In some cases, the desired performance decisions drew upon an agreed-upon assessment of risk, the existing capabilities of a building to withstand the motions of a pro-

jected event, and economic feasibility. Thus, the *Guidelines* documents focus and, in a sense, "discipline" rehabilitation decisions and the selection of target performance levels — from which then flow specific design choices, engineering parameters, and construction techniques.

Case 15: Santa Cruz, California

The city of Santa Cruz was heavily damaged by the 1989 Loma Prieta earthquake and faced a variety of reconstruction problems. A former city planner in Santa Cruz identified 25 post-earthquake challenges to his community, a full 18 of which are directly relevant to issues often encountered in the seismic rehabilitation of existing buildings foreseen by the Guidelines documents. Selected and slightly edited for use here, they are as follows:

- *The jurisdiction may have to add new administrative capacity (hire new staff), which involves both hiring time and learning time.*

- *Economic necessity may require more than simply rebuilding, especially when overlaid with new requirements for safety in retrofit and new construction. Retail trade may need to increase, and infrastructure upgrades may be required.*

- *Planning to rebuild accelerates attention to long-standing problems and issues (some of which will continue to prove intractable). Examples include defining appropriate levels of growth or economic development, upgrading of old infrastructure, and poor political environment (acrimonies, lack of inclusive decision-making processes).*

- *Rebuilding may require shifts in political and/or institutional patterns and habits.*

- *Political imperatives might be at odds with what makes sense from a planning or administrative perspective, which can make the decision-making process complicated and time-consuming.*

- *Special time and effort may be required to set up financial resources (tax measures, grant applications, redevelopment districts). Worse, resources may not be available.*

- *Decision-making may be delayed by the need to obtain information on and learn more about the* regional economic situation, financial options, development economics and potentials, geologic conditions, construction and design issues, and lender requirements.

- Political battles can command the time and attention of key actors and delay other decisions (e.g., historic preservation fights over buildings may delay decisions about adjacent properties and affect political discussion of other issues).

- New political interests may coalesce and need time to organize (e.g., a property owners association may become a necessity in an area where none existed previously).

- The local political system may have difficulty achieving agreement on key planning issues. Old adversaries may have to find common ground. Long-standing interjurisdictional disputes may have to be resolved.

- Philosophical differences may surface over the "proper roles" of the private and public sectors.

- New roles emerge. For example, property owners with no previous development experience suddenly become developers or a city with a reactive/regulatory orientation toward development may find itself having to solicit, if not court, new development.

- The most heavily affected areas may be the least economically viable parts of the community.

- Shortcuts are few. Legal and procedural requirements must be adhered to unless special legislation is pursued.

- Jurisdictions may have to seek, sponsor, or lobby for special state legislation.

- Perceptions of needs change, and planning may go in fits and starts.

- Organizing effective citizen participation is essential but takes time and effort.

- Displaced businesses and residents must be accommodated while long-term solutions are sought.

As this list makes clear, pre-earthquake and post-earthquake environments share many characteristics. The difference after a disaster, however, lies in a radically changed legal, regulatory, and political context — especially for seismic rehabilitation. After a major damaging earthquake, financial subsidies for repair and rehabilitation may suddenly become available, emergency authorities may be granted and exercised, and popular and media pressure to "do something" may emerge — all of which create the positive context for action only dreamed of by seismic safety proponents prior to the event.

In sum, earthquakes shoot seismic safety straight to the top of decision agendas, opening windows of opportunity for major advances. The question, of course, is how long those windows remain open before previous societal issues and problems regain their places on the agenda and new ones emerge, pushing seismic safety back down and starting the process all over again.

Perhaps of most direct importance for this discussion, damaging earthquakes may allow a jurisdiction that had been relying on simple attrition or following the lowest conflict model (voluntary) to move more aggressively on the earthquake-vulnerable buildings problem and utilize the "Informal/Encouragement Program" or go all the way to the formal "Mandatory Program."

Local economic conditions at the time of program enactment play a major role in seismic rehabilitation. For example, Los Angeles' Chapter 88 URM ordinance was passed in the "go-go" 1980s, a time of economic expansion and escalating property values, which made the financing of seismic rehabilitation projects easier.

Case 16: Portland and the State of Oregon

In 1993, western Oregon changed from Seismic Zone 2B to Zone 3 in recognition of new information about the risks of a subduction earthquake off the coast. This has had a significant impact on policies relating to existing buildings in that most of them now can be considered "dangerous buildings" because they were designed to a lower seismic standard.

In April 1995, the Portland City Council passed several ordinances that were developed by the Task Force on Seismic Strengthening of Existing Buildings. These constituted an interim policy that was to remain in effect until March 1997. The first ordinance took seismic loading out of the definition of dangerous buildings in the city's Dangerous Buildings Code. Other ordinances then codified several passive triggers that require seismic rehabilitation to current code or the suggested standard in the NEHRP Handbook for the Seismic Evaluation of Existing Buildings (FEMA 178), depending on the trigger. The following is a brief summary of the triggers:

• Changes of occupancy (to a higher standard based on UCBC ranking) and structural additions (that are not structurally independent) require rehabilitation to the current code standards.

• Alterations to most buildings valued at more than $100,000 require a FEMA 178 evaluation of the building. The data collected in this manner are to be used in developing the policies to be enacted after this interim period.

• Two types of alteration to URM buildings require rehabilitation to the FEMA 178 standard — reroofing (involving removal of the old roof or repair to more than 50 percent of the deck) requires anchorage of the roof system to the exterior walls and bracing of the parapets and alterations in a 2-year period that exceed $15 per square foot for the total net floor area trigger rehabilitation.

In 1995, the State of Oregon passed SB 1057 which created the Oregon Seismic Rehabilitation Task Force. The legislation directed the task force to provide recommendations to the legislature for its 1997 session. The task force has considered many of the topics important to any jurisdiction considering seismic rehabilitation programs including inventory data, mandatory and passive triggers, design standards, appeals, enforcement, liability, incentives, education and information, coordination and reporting, and needed legislation.

The task force filed its report on September 30, 1996. Legislation to begin implementation of the report was introduced in 1997 but it failed to pass. However, Oregon's legislature created the Oregon Seismic Safety Policy Advisory Council (OSSPAC). It expects to retain a focus on existing earthquake-vulnerable buildings as it considers long-term strategies.

Case 17: The Federal Case

In the 1990 re-authorizing legislation for the National Earthquake Hazards Reduction Program (NEHRP), Congress included a mandate that the President adopt "standards for assessing and enhancing the seismic safety of existing buildings constructed for or leased by the federal government." This one clause made the Executive Branch face the same issues that confronted so many private-sector building owners and local building officials — performance levels, priorities, scheduling, trigger mechanisms, funding, and others – but on a larger scale of course.

There was a very wide variance in cost estimates because of a lack of reliable data. The solution was therefore to adopt two parallel courses:

• Seismic rehabilitation is required for owned or leased buildings under a set of prescribed conditions ("triggers") when the upgrading of a building for other reasons will cost more than 50 percent of its replacement value and

• Collection of reliable cost data on which to base a more extensive, structured, and cost beneficial program of seismic rehabilitation also has started. In effect, this is a "Mandatory Program" model but one that is being implemented in an incremental and cautious manner pending the development of more reliable data on which to make such a significant public policy decision.

Implementation has begun. On December 1, 1994, President Clinton signed Executive Order 12941. This significant policy action, titled "Seismic Safety Of Existing Federally Owned Or Leased Buildings," established minimum seismic rehabilitation standards for "existing buildings constructed for or leased by the federal government which were designed and constructed without adequate seismic design and construction standards." While the Order establishes standards, a loophole is provided from what is an internal federal mandatory program. Under Section 3, "Implementation Responsibilities, federal departments and agencies are allowed to "request an exemption from this Order from the Director of the Office of Management and Budget." The conditions under which an exemption would be granted have not been defined, and no exemptions had been requested or approved at the time this publication was prepared. The results of this assessment could lead to a more active seismic rehabilitation program among federal agencies. Moreover, publicized upgrading of federal buildings in many communities might trigger greater attention to and action by local governments, building owners, and others with a stake in seismic rehabilitation.

BENEFIT-COST ANALYSES

Expenses associated with seismic rehabilitation are never trivial, largely because the basic structural frame of a building is at issue. In addition, many nonstructural and mechanical/electrical systems must be enhanced commensurately. Thus, the question of benefits justifying the costs keeps creeping into the discussions. Benefit-cost analysis can help overcome owners' initial resistance to investing in seismic rehabilitation in that it provides a structured way to compare the longer term benefits to be accrued when compared to the sometimes seemingly high initial costs.

Seismic rehabilitation costs money and money is scarce (by definition) but someone has to pay for it. In applying the *Guidelines*, a benefit-cost analysis is one way to link together and compare risk, expected building performance, estimated direct losses (including property damage, relocation costs, and losses in inventory, sales and rental income) with long-term benefits (the avoided future damage and ancillary losses) so that intelligent, or at least in formed, choices can be made about investing in rehabilitation. In the private sector, return on investment is another important factor that must be taken into account.

Case 18: The FEMA Benefit-Cost Modelling

FEMA has been addressing the fundamental "is it worth it" question since 1989 by supporting the development of basic benefit-cost methods, including manuals and software, that will help users analyze seismic rehabilitation possibilities. The models provide default values for key variables, but they explicitly urge users to provide ("plug in") more accurate and detailed local information whenever possible.

FEMA's initial efforts comprised two benefit-cost models for application primarily to privately owned buildings. The first focuses on single classes of buildings (e.g., URMs), and the second aggregates the results of several single classes to facilitate rehabilitation decisions about an entire area (e.g., Pioneer Square in Seattle or Old Sacramento in California). Additional cost data are contained in another FEMA document, NEHRP Guidelines for the Seismic Rehabilitation of Buildings: Example Applications (FEMA 276), expected to be available by mid-1998.

In essence, a benefit-cost analysis of the seismic rehabilitation of a building requires a cost estimate of the rehabilitation plan (always the easier part) and a probabilistic estimate of future benefits (more difficult). Benefits are calculated on a net present value basis to account for the time value of money. They also depend on the expected annual probabilities of future earthquakes and estimated "avoided losses." Those estimated avoided losses include building repair or replacement costs, damage to contents and inventory, relocation costs, lost income, and the monetary value of avoided deaths and injuries (based on a "statistical value of life"). The benefit-cost ratios tend to be high (favorable) when the building is of a hazardous class, the estimated cost of rehabilitation is modest, and the annual probability of earthquakes is high.

The appropriate FEMA publications and software are a pair of two-volume sets: A Benefit-Cost Model for the Seismic Rehabilitation of Buildings (FEMA 227 and 228, 1992) and Federal Buildings: A Benefit-Cost Model (FEMA 255 and 256, 1994) which also includes methods for estimating the value of public services.

In addition, a useful companion two-volume reference is available from FEMA — the second edition of Typical Costs for Seismic Rehabilitation of Buildings, Vol.1, and Supporting Documentation, Vol. 2. The new edition is based on a sample of 2,000 seismic rehabilitation projects throughout the country that were carefully screened and their cost data analyzed by sophisticated statistical techniques. In addition to mean cost figures, Volume 1 offers the user three optional methods of calculation, each yielding results that have variances that become smaller as knowledge about the basic characteristics of a single building or an inventory of buildings increases. Volume 2 provides the statistical underpinning of the data and information on additional costs associated with the nonstructural and administrative activities of a rehabilitation project. There already has been strong demand for these volumes, and their use is expected to grow considerably with time, especially as the implementation of Executive Order 12941, gains momentum.

In conducting benefit-cost analyses, it is important to recognize that rehabilitation costs can vary significantly. Such variations can be attributed to local economic conditions, prevailing wages, use of union or nonunion labor, times of day and days of week when work can be done, the extent of other upgrades required, the costs of finishes, and similar items familiar to those in the design and construction industries. In fact, the ancillary and "business interruption" costs of a major seismic rehabilitation project could actually exceed the direct costs of design, teardown, construction, permitting, etc. See Chapter 4 for an examination of potential societal issues by explaining the nature of each problem, typical issues that may need to be addressed, and various ways of solving each problem.

BUILDING OFFICIALS: THE EYE OF THE STORM

A jurisdiction's building officials are central under any of the three models and in any effort at seismic rehabilitation. Sooner or later they will be involved either actively or passively. To explain, a weather metaphor might be appropriate. Keeping in mind the increasing conflict potential in the three models, we can think of attrition as normal weather. The "Voluntary Program" is then a tropical depression and, the "Informal/Encouragement Program," a tropical storm. The "Mandatory Program" is a full blown hurricane. The building official is the constant, however, for he or she remains in the eye of the storm regardless of its size. In fairness, design professionals can become caught up as well.

Consistent with this perspective, a researcher once tried to contact the head of a building and safety department who was directing the preparation of a draft

hazardous structure abatement ordinance (i.e., this was a "Mandatory Program" case) and was taking an incredible amount of political heat as a result. Everybody was after him, and he was running from meeting to meeting. Not much can be done about the number of must-attend meetings for a building official involved in a "Mandatory Program," but one of the great virtues of the *Guidelines* documents is that, to return to the weather metaphor, these at least provide a sea anchor to the building official caught in the hurricane.

SPECIAL CIRCUMSTANCES AND WINDOWS OF OPPORTUNITY

It is almost a cliche to say that damaging earthquakes open "windows of opportunity" for advances in earthquake safety, but this is an actual truism for seismic rehabilitation. In California, still the perennial source for illustrations, in addition to code changes for new construction, both statewide and jurisdiction-specific seismic rehabilitation legislation came as direct results of various earthquakes from Long Beach 1933 through San Fernando 1971 to Northridge 1994.

While the *Guidelines* documents do not and are not intended to address the complicated issues involved in repairing earthquake-damaged buildings, pre-earthquake seismic rehabilitation of existing buildings and post-earthquake retrofitting of damaged buildings achieve the same purpose — lower risk to life and property. From a socioeconomic perspective, many of the same problems arise, and some wisdom can be exchanged. For any community considering seismic rehabilitation, the issue of what to require of new buildings always surfaces in discussions of what to require of existing ones. While the *Guidelines* documents offer several performance levels for rehabilitated buildings, many communities, especially those in lower risk seismic zones, will obviously be unlikely to apply to old buildings standards that exceed those required of new construction. Therefore, the core of an acceptable program may be correcting "fatal flaws" (those identified by the engineer and the building official) in various classes of existing buildings.

Chapter 4
TYPICAL SOCIETAL ISSUES IN SEISMIC REHABILITATION

Because rehabilitation deals with existing and usually occupied buildings, the range of socioeconomic issues likely to be encountered — and needing to be solved — can be formidable. Moreover, the intensity, nature, and complexity of such problems will vary somewhat from building to building even though sections or neighborhoods of cities and towns slated for seismic rehabilitation will have common problems depending on the demographic and socioeconomic characteristics of the designated areas.

This chapter breaks the overall forest of issues down into trees (at least the socioeconomic and administrative ones) that commonly arise in seismic rehabilitation programs. Each subject is discussed in terms of the nature of the problem, typical issues likely to arise in connection with that problem, and some possible ways to solve or at least ameliorate the negative impacts of the problem. It is an axiom that the lower the level of conflict, the easier it is to first adopt and then implement measures that have retroactive characteristics.

The first section of this chapter discusses demographic, social, and economic factors while the second section treats public policy and administrative issues typically involved in seismic rehabilitation. For example, ownership patterns, income levels, historic properties, and occupancy characteristics are contained in the first section while policy formulation and adoption strategies and legal and program management issues are included in the second section.

An overriding concern in seismic rehabilitation has to do with accommodating the building's intended use. Obviously, all design professionals know they have to accommodate the owner's intended uses of the candidate building. However, seismic rehabilitation projects often are technically tricky and part of their success depends on achieving an effective balance between improved earthquake safety and func-

tionality. A related FEMA publication (FEMA 172, p.17) notes that:

> Most buildings are intended to serve one or more functional purposes (e.g., to provide housing or to enclose a commercial or industrial activity). Since the functional requirements are essential to the effective use of the building, extreme care must be exercised in the planning and design of structural modifications to ensure that the modifications will not seriously impair the functional use.

DEMOGRAPHIC, SOCIAL, AND ECONOMIC FACTORS IN SEISMIC REHABILITATION

Because existing buildings were built to earlier standards and often are occupied, a wide spectrum of social and economic problems may be encountered when seismic rehabilitation is considered. Some or all of them may arise during the project planning process. The most significant topics are discussed below: the distribution of impacts on various segments of the community; means to minimize business interruption, occupancy dislocation, and the loss of housing; the treatment of historic properties; and approaches for financing seismic rehabilitation. For example, when San Francisco examined socioeconomic factors related to its URM buildings, it found that 7 percent of the businesses were in URMs, 7.5 percent of jobs were in URMs, and 7 percent of the URMs provided housing, even though only 3.7 percent of the city's residents lived in URMs.

Evaluating the Distribution of Impacts Due to Seismic Rehabilitation

Nature of the Problem: Seismic rehabilitation affects people differently. There are organized interests that may become mobilized, and there are latent ones that may emerge during the process of formulating seismic rehabilitation policy as well as around specific projects. Chambers of commerce, merchants associations, local design professionals, and boards of realtors are examples of formal interests while building owners, loosely structured neighborhood groups, or even tenants within individual structures may organize around a given project.

It seems clear that supporters of seismic rehabilitation may be a coalition of local and distant design professionals, building officials, and others committed to seismic rehabilitation, but the opponents most often are totally local, those whose immediate interests are most likely to be directly affected. It is important, therefore, to anticipate the composition and range of interests of the coalitions that might form and to evaluate what the impacts will be on each and how each will perceive and therefore react to proposed seismic rehabilitation programs and projects.

Typical Issues: Several key issues will arise in virtually every seismic rehabilitation policy development process:

What is the scope of the seismic rehabilitation effort? It matters greatly if the project is one building, a well defined portion of the city (e.g., "Pioneer Square"), a concentrated or evenly widely distributed class of existing buildings (e.g., URM bearing wall structures), or a targeted use (e.g., theaters and churches). The scope of the seismic rehabilitation program will define the interests most likely to become involved in the process.

What existing local groups are likely to become involved, and what will be their particular interests in seismic rehabilitation?

Can support or opposition be expected from latent interests that might define seismic rehabilitation as an issue?

What work will be required, how much will it cost, and when must it be completed?

The answers to these questions define the potential intensity of the interests' positions.

Solving the Problem: Several actions can be taken to anticipate the impacts of and the interests likely to be affected by seismic rehabilitation projects and programs. Some suggestions include:

Identify government agencies, community groups, and professional and business associations that historically have played key roles in planning and zoning, redevelopment, building code, housing, and related issues. This information often can be obtained from local agencies. Review the positions taken and attitudes expressed by these groups on related issues.

Identify latent or emergent groups that may or may not have been actively involved in the past but that could become so depending on the focus of the seismic rehabilitation program. This may be more difficult than identifying formal groups, but it is worth the effort because unexpected vocal opposition, even from a small but highly visible group, can have serious consequences for proposed projects.

Hold well announced community meetings to introduce the concept while the program is still in the formative stage. One effective mechanism is to then form a "Community Advisory Committee" whose members represent all interests. This group then can examine the issues in a common framework and perhaps reach consensus on critical issues. Community meetings and advisory groups require extensive technical and staff support, and this workload should be anticipated.

Inform the local media, especially the local newspapers that tend to follow local issues for extended periods and that can have a major influence on the acceptability of seismic rehabilitation programs. This takes skill and preparation, but the evidence is clear that newspaper support is very important and that newspaper opposition can prove fatal. Skillful work with the media may even prevent seismic rehabilitation from becoming a "hot" issue.

Determining Occupant Dislocation and Business Interruption

Nature of the Problem: While extensive seismic rehabilitation projects do not always, they can require relocation of building owners, employees, commercial tenants, and residents. If the construction work is relatively minor but cannot be accomplished with the occupants in place (during off hours when the building is closed), it is better to face this issue as early as possible and allow plenty of time to solve it. If the seismic rehabilitation project involves leased space and if it is encumbered with a mortgage, loss of rental income to service the debt can become a major concern. It is therefore important to anticipate how potential extra direct costs and inconveniences can be ameliorated in the quest for safer buildings.

Typical Issues: While only some of the impacts are financial, they are the major ones. Typical issues within this context include:

How feasible is it to perform the seismic rehabilitation work without having to relocate the occupants to other locations? This depends a great deal on the building's occupancy and some — even extensive — seismic rehabilitation projects have been completed without relocation.

In addition to the costs of construction, how can the owners continue to pay the mortgage, insurance, taxes, and other operating costs when the building is not generating income? Unless owned outright with costs financed from savings or from a capital improvement pool of the building owner, this "cash flow" question becomes important.

Who is responsible for notifying the tenants and residents, paying the costs of relocation, and allowing sufficient time for the relocation process to occur? These issues are at the heart of the viability of commercial, residential or business occupancies. The answers often depend on the availability of other nearby comparable space, equitable rents, and the use of various subsidies.

Solving the Problem: A variety of actions can be taken to ameliorate these problems including the following:

Ensure that the initial feasibility study of a particular seismic rehabilitation project can address the question of whether the work can be done without substantially disrupting operations. It is much easier in single occupant office buildings or commercial properties that are empty during the late hours and where some internal temporary space-sharing can occur than in multiple tenant or residential occupancies. In addition, the contractor will have to carefully ensure that the construction work areas are sealed adequately and that time is allowed for thorough clean-up before normal business operations resume. One also must be aware of other problems (the exis-

tence of asbestos) that could make seismic rehabilitation more complex and expensive.

Cash flow for debt service and operating expenses is critical. Anything, including seismic rehabilitation, that interrupts that flow can have major implications. Nevertheless, the situation will vary with each case. Internal operating or capital improvement monies could be used where they exist and rehabilitation is included in scheduled outlays. As incentives, local governments could suspend property taxes and other charges until the building is ready to be reoccupied. Other types of remodeling and rehabilitation often are done upon transfer of the property to new owners or when major tenants relocate to other facilities. Large tenant commercial leases often last for about five years, and rehabilitation could be scheduled to coincide with a tenant's decision not to renew its lease. Financial advisors to both owners and local governments may well be aware of other possibilities to soften the cash flow impacts of seismic rehabilitation.

The picture is less clear for commercial lessees and residential renters. The minimum is to provide as much advance notice as possible so they can take appropriate steps to minimize the negative impacts. One possible strategy to ameliorate the costs to such occupants could be to help them find temporary and comparably priced nearby space coupled with giving them "first right of refusal" to return to the rehabilitated building. Local governments may be able to offer other incentives through neighborhood revitalization and community redevelopment measures. Such techniques often involve tax, loan, and other incentives, and they can include relocation services assistance.

Minimizing the Social and Economic Impacts on Housing

Nature of the Problem: Although a relative term in any economic setting, "affordable housing" deserves a special focus because of its importance to the community, lower income neighbors, and social justice. Sadly, in many communities it often is the lower income and, just as often, non-English speaking unorganized members that also reside in the more earthquake-vulnerable buildings. When displaced by damaging earthquakes, these same people also become the most dependent on emergency shelter, financial assistance, and other direct aid. The more affluent find temporary quarters, have other financial resources, and generally are better able to adjust.

Recent research (Comerio, 1995) based on data about the housing losses from the 1994 Northridge earthquake estimates that 60,000 dwelling units could be "significantly damaged" after a major event in the region. Of these 60,000, only 7,000 would be single-family dwellings. Thus, about 53,000 units would be apartment units and about 50 percent would have to be vacated because of the damage. Using 3.5 persons per apartment unit as an average, this means that over 90,000 renters could be homeless. A comparable calculation for an equivalent earthquake on the San Francisco Bay area's Hayward Fault is more depressing because of higher population densities. About 240,000 housing units could be significantly damaged, of which about 100,000 could be unoccupiable. Using the same 3.5 person household average, the homeless could number about 350,000 people (Comerio, personal communication, September 1995). Although less glamorous, technically challenging or financially rewarding than other forms of seismic rehabilitation, the need for effective mitigation measures to protect the nation's housing stock is great.

Typical Issues: While the major issues are comparable to the earlier ones, the main difference is that housing rehabilitation focuses on small economic units (individuals and families). Consequently, it is important to determine:

How long will the project take and where can the occupants go for the duration of the work?

Can the owner afford the rehabilitation work and are there any incentives or cost offsets that can help pay the costs?

If the occupants are renters, will they be able to afford the rent of the rehabilitated housing unit?

If the occupants are in poor health or disabled and have to be relocated, can support be provided in the new locations?

Will the owner demolish the building and put occupants on the street?

Will the owner remove housing units on the site and use the building for something else?

Solving the Problem: Generally speaking, more affluent residents can afford to pay for and vacate their housing during substantial remodeling and rehabilitation. As income declines, however, this easy option disappears. Thus:

Fortunately, even in the smaller (1 to 2 story) single- and multiple-family units, many housing rehabilitation techniques can be employed without requiring occupant relocation. Examples include bolting foundations to sills, tying chimneys to the structure, installing effective shear walls, and applying other sound and well understood techniques. Moreover, such work can be linked to other changes being made to the units. Depending on the scope, such work often lasts only a few days or weeks. However, the seismic rehabilitation of larger buildings, (e.g., apartment buildings) can become complex, costly, and time consuming. Such work is comparable to rehabilitating commercial structures and many of the problems will be the same. Condominiums and other "planned unit developments" create special problems because of the joint maintenance responsibilities for the common areas and governing processes involved in managing such developments.

The affordability of seismic rehabilitation is a function of the financial resources available and that depends to a great extent on whether or not the building is owner-occupied. While desirable, there are very few financial incentives available to housing owners to stimulate seismic rehabilitation. This remains one of the major challenges to speeding up the process. Some aids do exist. For example, California law prevents the raising of property taxes when seismic safety improvements are made to buildings so at least the owner is not penalized by a tax increase. The popular equity lines of credit can be used for home improvements and the interest is tax deductible. Savings also can be used.

Increased rents often are a result of building rehabilitation. Covering the costs of rehabilitation and attracting a more affluent clientele are frequently interwoven motives along with a desire to increase the market value of the structure. This creates special problems for lower income renters. Some techniques for minimizing the impact of higher rents include:

local officials giving higher priority to people displaced by seismic rehabilitation and qualifying them for rental assistance programs; increasing other cost offsets such as providing renters with free or reduced- cost public transportation vouchers and other benefits; and allowing the adjustment of rents within specified time and monetary limits. Nevertheless, the fundamental tension will continue between achieving a safer building (a public good) and controlling the cost of living (a private matter). The extent to which seismic rehabilitation can be directly or indirectly subsidized can greatly affect the continued availability of affordable housing.

Historic Properties Destined for Seismic Rehabilitation

Nature of the Problem: During the past 20 or so years, efforts have been mounted to identify, preserve, and tightly control the uses of and modifications to properties considered "historical." Seismic rehabilitation work on buildings falling into this category can be very challenging for the design and construction community because of special regulations, the existence of delicate finishes and archaic (and often mixed) materials, aesthetic needs, and little or no information about the site, foundation or structural conditions of the structure. Whenever historic buildings are involved, it is very important to carefully review governing codes, standards, and other applicable materials such as the Secretary of the Interior's *Standards for Rehabilitation and Guidelines for Rehabilitating Historic Buildings* (see Chapter 6).

One structural engineer experienced in the seismic rehabilitation of older and historic structures noted that (FEMA 237, p. 77): "All of these [archaic] systems were designed prior to the development of seismic standards for buildings. Probably none were designed for seismic performance at all." However, because such buildings are intended to be "permanent" fixtures of the built environment, they merit seismic rehabilitation. Nonetheless, ". . . in any community the presence of even a few historic buildings will greatly complicate the implementation of either voluntary or mandatory seismic protection policies for existing buildings" and the ". . . effort to extensively strengthen the building can tend to result in the removal of much of the original material, the obscur-

ing of original features, or the introduction of visible bracing elements. . . ."

On the other hand, the Preservation Tax Incentives for Historic Buildings have provided the means for rehabilitating many buildings. The initiative allows a 20 percent investment tax credit (ITC) for the certified rehabilitation of an income-producing, depreciable certified historic building and a 10 percent ITC for the rehabilitation of income-producing, depreciable buildings (excluding residential rental) built before 1936. Seldom does the seismic rehabilitation cost more than the 20 percent ITC.

Typical Issues: From our perspective, a number of issues related to the seismic rehabilitation of historic buildings are important including:

What is an historic building? To quote from an earlier FEMA document (FEMA 237, p. 79):

> . . . there is no indisputable definition of "historic building." Guidance is provided on rehabilitation of historic buildings in state documents such as the State Historic Building Code in California or in federal documents such as the Secretary of the Interior's Standards for Rehabilitation and Guidelines for Rehabilitating Historic Building and associated guidance. Buildings may be listed on the National Register of Historic Places, a state historic register, or a local listing that has official status. In some cases, rather than a simple determination that a building is on or off such a list, a ranking of the degree to which a building is historic is made with reference to a local priority or historic value scale. Criteria and the process for placing buildings on such lists vary and can be influenced by local demands that include considerations beyond this historic quality of an individual building, such as desires to minimize density and land use changes or to avoid renovation or new construction that would introduce higher rents.

Chapter 1 of the *Guidelines* volume, however, states that:

> It must be determined early in the process whether a building is "historic." A building is historic if it is at least 50 years old and is listed or potentially eligible for the National Register of Historic Places and/or a state or local register as an individual structure or as a contributing structure in a district. Structures less than 50 years old may also be historic if they possess exceptional significance. For

historic buildings, develop and evaluate alternative solutions as to their effect on the loss of historic character and fabric, using the Secretary of the Interior's *Standards for Rehabilitation.*

Who has jurisdiction of the building? This seemingly simple issue is a very important one for owners of historic buildings that are candidates for seismic rehabilitation. One needs to determine who actually owns the building (e.g., private party, charitable or nonprofit organization, foundation, or government agency). It also is important to determine who has jurisdiction over the building (local, state, or federal government) and, consequently, which codes or regulations will apply to the rehabilitation project. For example, the city of Seattle has jurisdiction over every publicly or privately owned building except those that belong to the federal government. While not all states may have a state historical building code, the city of Seattle enforces the State of Washington code. Moreover, the owner and his/her design professionals may have to observe other requirements depending upon which category or register the historic building appears is listed on. This specialized field requires specialized expertise.

What is the occupancy and the amount of operational disruption that can be accepted during construction? Some historic buildings, like George Washington's home in Mount Vernon, are landmarks open to visitors while others, such as California's restored State Capitol, function as full-time office buildings and house key activities and records. At the local level, some historic buildings are in older commercial areas of once small towns and their activities are important to the economy of the area and the businesses or residents housed there. In these cases, the amount of disruption, the need for relocation, the nearby availability of affordable alternative space, and the scheduling of the work become important considerations.

What level of performance is desired and how much will it cost? While key questions for all buildings, they are especially important for historic structures because the answers tie back to the building's importance, replacement cost (if it can be replaced at all), the objective earthquake risk, acceptable levels of damage, types of historic finishes, and sources of funding.

Solving the Problem: Dealing with the unique problems posed by historic building rehabilitation can take several forms, alone or in combination depending on the circumstances. Owners sometimes have relatively little to say about what can be done to their designated historic buildings. Therefore, suggested strategies include:

Determine if the particular building has indeed been designated historic and by whom. This information will determine whose design and construction regulations and enforcement processes will govern the project.

Review the regulations and processes, paying particular attention to any special standards or exemptions, design review requirements, appeals or approval processes, flexibility in time for compliance, alternative approaches, and similar factors.

Like other buildings, determine the current use of the historic structure and what the dislocation and other extra needs might be to accommodate the occupants and functions. This will require some effort if these problems can be handled imaginatively, easily, in a timely fashion, and affordably.

Analyze the exposure of the building to the expected earthquake risk in the region and balance this with the building's value to the community. There is the need to judge the building's long-term significance, its occupancy and function, the cost to replace it versus the cost to repair it occasionally, and other factors. The answer will almost never be clear. Given the desired permanence of historic buildings, it may mean that the rehabilitation decision will have to consider lower probability but more severe ground motions and more earthquake occurrences during its estimated post-rehabilitated lifetime.

Select the desired seismic rehabilitation performance level from the Guidelines. As with other buildings, this is critical because the selection will drive the design alternatives, costs, and scheduling. FEMA 237 (p. 80) notes that such an ". . . approach will help preserve historic buildings from earthquakes, even if they are strengthened only up to a minimum life-safety level, and prevent the situation from developing where the historic buildings will be the most hazardous in a community."

Determine what efforts are needed to accommodate the relocation of the occupants, time needed for rehabilitation, and how and if the most important functions performed in the building can be or need to be maintained. Solutions to these issues will vary with each project.

Involve and, to the extent possible, obtain consensus among the controlling stakeholders that the preferred seismic rehabilitation technique will be effective and workable. Historic buildings are highly visible and the foci of often influential advocacy groups. Therefore, it is important that advocates be informed of the potential project and be brought into the process early; it is worth the up-front investment of time and energy.

Obtain the advice of state historic preservation officers and other specialists in the preservation of historic finishes and involve them from the very beginning of the rehabilitation process.

Finding ways to address the unique problems associated with the seismic rehabilitation of historic buildings will help ensure that the threat of earthquake damage to these structures will be reduced and that they will continue to be important reminders of earlier times and events.

Financing Seismic Rehabilitation

Nature of the Problem: While regular building maintenance is a continuing operating cost, seismic rehabilitation and other major capital improvements can be expensive, especially for larger buildings. The ability to finance such improvements varies greatly with the owner's ability to pay, what seismic rehabilitation work needs to be done to the building, and what other improvements will be made at the same time. Since each building has its own story, it is very important to determine if the costs of seismic rehabilitation are affordable. One observer noted that, especially in the eastern United States, most older buildings have expended much of their useful life and frequently may not be providing adequate financial returns in their current condition. Many engineers have submitted reports about what should be done to a building to improve its earthquake performance, only to see little or no subsequent action taken.

It is clear, however, that the pace of seismic rehabilitation is increasing in places like California where frequent recent events have occurred; higher risk is perceived; and lenders and insurers are evaluating properties more closely, limiting coverages, raising deductibles, and taking other measures to lessen their exposure to earthquake losses.

Typical Issues: Successfully answering several questions is at the heart of investing in seismic rehabilitation. Savings, loans, operating revenues, or capital improvement funds are traditional and usually private-sector sources of money to finance seismic rehabilitation. However, some may ask:

Are there government programs available to help pay for seismic rehabilitation?

What incentives exist that at least could help offset the direct costs of seismic rehabilitation?

Can an owner adjust his/her insurance costs to free up funds for seismic rehabilitation?

Solving the Problem: The financing mix necessary to increase the earthquake resistance of existing buildings will vary on a case-by-case basis, but some suggestions can be provided:

If a public agency, the owner can seek direct appropriations through the normal budgetary process. Other possibilities include raising money through the issuance of bonds and other forms of financial participation in public projects. For example, in 1990, California's voters approved Proposition 122, which made $300 million available to strengthen existing buildings owned by state and local governments. Soon after the 1971 San Fernando earthquake, the U.S. Department of Veterans Affairs secured funding through the regular budget process to seismically evaluate and rehabilitate many of its older buildings across the country. As noted earlier, the school district in Clayton, Missouri, raised money via a bond issue and San Leandro used Certificates of Participation.

Limited incentives (mostly indirect) exist and should at least be considered as ways to offset the direct costs of seismic rehabilitation. In 1990, California's voters approved Proposition 127, which exempted seismic rehabilitation improvements to buildings from being reassessed to increase property taxes.

Special funding and tax measures often are part of community redevelopment programs and seismic rehabilitation costs could be considered eligible project costs. State legislation might be needed to expand the definition of "blight" to include hazardous buildings. Bonds might be used to guarantee loans for rehabilitation, but this may be a problem (as it has been in California) because bond holders take precedence over mortgage holders in the event of foreclosure and revenue bonds must be repaid from income generated by the projects they fund.

While mobile homes are not "buildings" in *Guidelines* terms, San Bernardino County, California, is implementing a financial incentive program to seismically strengthen these structures. Learning from the over 7,000 mobile homes damaged in the Northridge earthquake, the county has selected a manufacturer of a foundation bracing system. Owners of existing units must use this approved system to qualify for a low-interest loan program. It is financed by a taxable 7-year bond issue, and the bond buyers receive 10.25 interest. Described as a "win-win" situation, it is "revenue neutral" to the county and participating cities. In addition, low-income mobile home owners also may be eligible for redevelopment funds and other federal and state assistance (CSSC, September 1995)

In the city of Berkeley, 50 of the property transfer fee is waived when a new owner of a house bolts it properly to the foundation. San Leandro, California, waives the need for a building permit and its fees when an owner uses standard guidance provided by the building department to secure his or her home to its foundation. San Francisco's $350 million bond issue (Earthquake Loan Bond Program, November 1992) designates two-thirds of the money ($233.3 million) for the seismic rehabilitation of housing. This means that owners get lower interest rates (about 1.5 below the bank's rates) and better lending terms if the rents are kept affordable. Loans to seismically rehabilitate housing units under this program were costing only about 3 percent in the fall of 1995.

Other types of incentives have been discussed or used in a variety of different contexts. Point-of-sale disclosure requirements and inspections of and repairs to specified conditions or items could be required for residential and commercial properties.

Post-disaster aid might be allocated in ways that reward those who invested in seismic rehabilitation rather than those who did not.

Some post-earthquake assistance measures might be adapted to act as pre-earthquake seismic rehabilitation incentives. For example, in addition to waiving permit fees to help recover from the Northridge earthquake, Los Angeles waived sewer connection and business relocation permit fees and extended the payment schedule for business taxes for six months. The city loaned victims hundreds of millions of dollars as "loans of last resort" to help repair damaged housing. Business assistance centers were set up to help small businesses prepare loan applications and supporting business plans. The housing department hired "work out loan specialists" to help design loan packages and solutions and also to become sales people who contacted individual property owners to convince them to apply. Some damaged commercial properties are being taken over by nonprofit organizations, which entitles such organizations to various assistance programs and incentives not available to private owners.

The underlying principle, however, is that the mix of incentives must support the goal of seismic rehabilitation and be consistent with state, local, and private financial laws and practices in the area. The property insurance industry, especially after experiencing major losses in recent years, is becoming more active in the field of mitigation, and seismic rehabilitation is one area of interest. Perhaps this will lead to rate differentials (incentives or disincentives) for at least high value properties where seismic rehabilitation work is accomplished.

Risk managers for some private owners have assumed more of the exposure by changing the mix between premiums, deductibles, and self-insurance reserves, which has sometimes freed cash for seismic rehabilitation. The objectives are not only to protect the physical plant but to lessen the business interruption costs. As premiums and deductibles have increased and property insurance carriers have placed limits on how much they will pay the policyholder, such strategies have become more common. In lieu of paying higher premiums, one approach is to pay for seismic rehabilitation from savings achieved by taking lower coverages and assuming higher deductibles. Some organizations have even established special reserve accounts to have cash available to make early repairs to damaged buildings. This risk management practice also has been followed by some government agencies whose continued operations are of critical economic importance (e.g., port authorities). While some seismic rehabilitation work can be undertaken with these funds, such special "force accounts" basically provide ready cash for post-earthquake emergency repairs and mitigation actions, even though the entities involved probably will qualify for later federal disaster assistance payments.

PUBLIC POLICY/ADMINISTRATIVE ISSUES IN SEISMIC REHABILITATION

Important policy and administrative issues are inherent in the process when local and state governments exercise their powers and become involved in seismic rehabilitation programs (even though they also may arise occasionally in voluntary efforts). This section focuses on factors that might "trigger" seismic rehabilitation, local capabilities to regulate and perform such work, managing the political issues in program adoption and implementation, addressing common legal problems, choosing which buildings (or how many) to rehabilitate, evaluating the local fiscal effects of rehabilitation, and achieving the mitigation of other hazards while reducing seismic risk.

Triggering Seismic Rehabilitation

Nature of the Problem: Much of the information in the *Guidelines* documents eventually could be used to develop formal seismic rehabilitation codes and standards for use by state and local jurisdictions. Often the rehabilitation of existing buildings requires that permits be obtained, plans be approved, and inspections be conducted. Design professionals and building officials are aware that the extent of a proposed remodel often "triggers" requirements to upgrade the building in many ways. Therefore, one key local policy decision involves determining if and under what circumstances seismic rehabilitation standards or requirements become a required

(triggered) part of a more extensive renovation or remodeling project.

Triggers fall into two principal categories — active and passive. Active ones are instigated by building departments and include such things as ordinances requiring the seismic rehabilitation of nonductile concrete frame buildings, the securing of parapets on URM buildings, or the replacement of damaged structural members with those that meet current requirements. Passive triggers are those that come into play when a building owner proposes to make changes to the structure, use or occupancy of the building, when vacant buildings are to be reoccupied (especially when deterioration is evident), and when the owner proposes to sell the building and the transaction is governed by disclosure requirements. Some common triggers are activated if a building:

- Is in a defined class (e.g., URM, pre-1973 tilt-up)?

- Is proposed to undergo major remodeling, (e.g., costing more than a specified amount or 50 percent of its replacement value)?

- Will have a major increase in the number of occupants (e.g., warehouse to offices)?

- Will change uses (e.g., manufacturing to trendy loft-style apartments)?

- Will be changing owners under certain circumstances?

- Is located in a special district (e.g., San Diego's Gaslamp Quarter)?

While triggers are technical matters, they are not discussed in the *Guidelines* documents because their selection is a fundamental policy choice in seismic rehabilitation. Triggers may not specify what the extent of work must be, but they do function as an "off-on" switch.

Typical Issues: Several key questions should be addressed in deciding whether or not to use major remodeling as a trigger for seismic rehabilitation and, if yes, what the specifications should be. Some questions include:

Should triggers be included in a negotiated or formally mandated program at all or should seismic rehabilitation be left to the judgement of the parties involved? Examples of both approaches exist. A traditional approach is that when the total project cost amounts to 50 percent of the replacement value of the building in question, the local building code requires that other modifications be made or that it meet the requirements for new buildings. This has the advantage of being clear to the parties involved (i.e., the rules of the game are known). While trigger requirements are important parts of the building regulatory environment, experience has shown that projects sometimes are broken down into discrete smaller projects so that triggers and other process requirements are avoided. This incremental approach to rehabilitation may achieve a narrow set of owner-preferred property improvement objectives, but it can miss important public safety objectives. Another approach allows the building official to determine when seismic rehabilitation will be required for a project. When it is, the owner, the involved design professionals, and the building official negotiate the nature and extent of the seismic rehabilitation work on a building-by-building basis.

What should be the rehabilitation standard? Concern is frequently expressed that a rehabilitated building must meet the local code's seismic requirements for new buildings. While it is especially important to increase the capacity of a structure to resist earthquakes, it may not be feasible to require conformance with standards for new buildings for design, cost, practical or political reasons,. Some seismic improvement is better than none.

If seismic rehabilitation is triggered and the project goes forward, should the owner be guaranteed that further and future retroactive requirements will not be demanded for some specified time? Seismic rehabilitation often is expensive. It is important, therefore, that owners be granted some "grandfather" guarantee that further seismic and possibly other upgrades will not be required for some specified (preferably lengthy) period of time.

Will the proposed seismic rehabilitation project trigger other requirements that, when taken together, result in a too complex or expensive project? Typical requirements include hazardous material (asbestos) remediation, access for the disabled, and the installation of fire protection sprinkler systems. While each has an important purpose, it may be possible to

establish a seismic rehabilitation program to minimize the triggering of these other requirements. For example, San Francisco's building code regarding the seismic rehabilitation of URM buildings provides owners with an opportunity to obtain an exemption from disabled-access requirements if the work is less than about $86,000 (adjusted for 1996) based on "hardship" or "legal and/or physical constraints"; requests for exemptions are handled by an access appeals board.

Solving the Problem: The key to solving the problem of whether or not to include seismic rehabilitation triggers for major remodeling is directly related to the fundamental policy choice the community makes to achieve seismic safety in existing buildings. If the choice is to formally require seismic rehabilitation, the remodeling program should contain clear statements about the criteria that will trigger seismic rehabilitation requirements. However, if the informal/encouragement approach is used, the local building official has much greater latitude.

If triggers are to be formally prescribed, then choices will have to be made about what they are. In general, a "trigger" reflects a central policy decision for it determines when a building is or is not subject to seismic rehabilitation requirements. The choice of triggers is, therefore, at the crux of the seismic rehabilitation policy formulation and adoption process.

The standards governing existing federal government buildings (ICSSC, RP4, p. 7) specify that a building shall be evaluated and unacceptable risks mitigated when any of the following triggers occur:

- A change in the building's function occurs that results in a significant increase in the building's level of use, importance, or occupancy as determined by the federal agency;

- A project is planned that will significantly extend the building's useful life through alterations or repairs that total more than 50 percent of the replacement value of the facility;

- The building or part of the building has been damaged by fire, wind, earthquake, or other causes to the extent that, in the judgment of the federal agency, structural degradation of the building's vertical or lateral load-carrying systems has occurred;

- The building is deemed by the agency to be an exceptionally high risk to occupants or the public at large; or,

- The building is added to the federal inventory through purchase or donation after the standards were adopted for use by the federal government.

Triggers, however, can be narrowly defined so as to severely limit seismic rehabilitation. A Utah state law that became effective on January 1, 1993, requires that all commercial buildings built before 1975 be evaluated for seismic hazards and that corrective actions be recommended by the evaluating engineer. However, as a state newsletter noted, the law has been largely ineffective because it is triggered only "when said building is undergoing reroofing or alteration of or repair to" parapets and other such limited items (State of Utah, p.5). The difficulty is compounded by building officials being unaware of the change or by owners contracting for reroofing without obtaining a permit.

While less formal than the triggers discussed above, there are other mechanisms ("pseudo-triggers") that can help achieve limited forms of partial or incremental seismic rehabilitation. Studies performed by Building Technology, Inc., (1994, p. 1) on how to improve the seismic safety of existing school buildings in several states focused on linking "incremental seismic retrofit (rehabilitation) opportunities to specific maintenance and capital improvement projects." For example, roofing maintenance and repair could include anchoring of parapets or roof-mounted equipment and shear walls could be strengthened with plywood when finishes are exposed or removed for other reasons.

Assessing Design, Regulatory, and Construction Capabilities

Nature of the Problem: The rehabilitation of existing buildings challenges all involved parties — architects, engineers, other design professionals, local planning and code enforcement officials, the myriad of construction trades, and the owners. The challenges are especially acute for seismic rehabilitation because the requisite knowledge, experience, and capabilities vary widely across the United States.

Even in California, where the number of people technically qualified for seismic rehabilitation work is comparatively large, the pool is still quite shallow. Clearly, a successful seismic rehabilitation project depends directly upon the knowledge and experience of those involved. This suggests that anyone initiating or regulating a rehabilitation project with a seismic component should not only carefully evaluate the technical qualifications of those involved but should also be prepared to supplement or require additions to a rehabilitation team.

Typical Issues: To determine if adequate technical, regulatory, and construction experience and knowledge are being applied to a seismic rehabilitation project, several questions must be asked:

From a design and construction perspective, how complicated is the project and is the project team fully qualified to perform the specific work proposed? Although every building has its own story, some types or classes of structure are simpler to rehabilitate than others. Unique or complex structures are especially problematic to rehabilitate, and while substantial documentation and rehabilitation experience exist for some structure classes (e.g., URM bearing wall and tilt-up buildings), considerably less documentation and experience are available to guide the rehabilitation of other kinds of construction.

Whether seismic rehabilitation is just one part of or is the principal reason for a project, the earthquake engineering qualifications and experience of the project team become very important considerations. Ensuring that the proper expertise is applied to the project goes a long way toward effective quality control throughout the process. Careful design is the first part of a rehabilitation process; adherence to that design during the actual work is the second part. Both are important.

When seismic rehabilitation projects are few and far between and when no prescribed guidelines or standards exist, how can the responsible building official be confident that he or she has the technical competence available to ensure that the seismic rehabilitation work is adequately planned and properly performed? Given the unusually high degree of judgment involved in seismic rehabilitation projects, it is important that the local regulatory agency have knowledgeable and experienced expertise available either on staff or externally.

Where can additional seismic rehabilitation design and construction expertise and capabilities be obtained? The securing of such expertise is a major concern in every project, but it is even more of a problem in areas where comparatively little experience exists and where the practicing architectural, engineering, and construction communities are less well informed about earthquake engineering and seismic rehabilitation. In these situations, local building rehabilitation capabilities must be directly supplemented with specialized earthquake-related knowledge.

Solving the Problem: Many individuals, especially from lower risk seismic zones of the United States who helped design Chapter 5's Applications Scenarios, raised all of the preceding questions. They were clearly concerned about the adequacy of the design, engineering, construction, and regulatory capacities in their locales to successfully perform seismic rehabilitation projects. A few suggestions are offered:

The Guidelines documents provide, for the first time, comprehensive reference information for design professionals to use in strengthening seismically weak buildings. These documents reflect the state of knowledge and practice that existed at the time of publication. While each building has its own story and despite limited experience with the performance of seismically rehabilitated buildings in actual earthquakes, the *Guidelines* documents provide a reasonable basis for undertaking such projects.

Professional societies and trade groups (including local and state architectural and engineering organizations, contractors associations, and builders associations) are often helpful in locating members with seismic rehabilitation experience. Such national organizations as the Earthquake Engineering Research Institute (EERI) in Oakland, California also can help as can such university-based research organizations as the National Center for Earthquake Engineering Research (NCEER) at the State University of New York (Buffalo campus), the Earthquake Engineering Research Center (EERC) at the University of California at Berkeley, and the John A. Blume Earthquake Engineering Center (ERC) Stanford University.

If time allows, an individual can increase his/her expertise by self-study and by attending technical meetings and seminars conducted by a variety of entities. Peer contacts also can be an efficient way of locating appropriate consulting assistance. If sufficient long-term seismic rehabilitation work can be expected, adding expertise directly to the staffs of design, engineering, construction, and regulatory organizations is another possibility. Indeed, for practitioners, adding such expertise might prove a competitive advantage in their market areas.

Depending upon the project and situation, a variety of ad hoc mechanisms such as arranging for independent reviews by other (fully capable) practitioners can be used during seismic rehabilitation projects. Other such mechanisms include forming project-specific panels of expert reviewers and, in the case of regulatory agencies, establishing appeals boards to advise on or even approve seismic rehabilitation projects. The latter mechanism is especially helpful if no formal standards exist or if the project's complexity requires substantial judgment and discussion.

Managing the Program Model's Adoption and Implementation Processes

Nature of the Problem: As noted in Chapter 2, the "Mandatory Program" can be the most controversial to enact and implement, primarily because it requires formal action by such elected bodies as town councils and boards of supervisors or commissioners. By necessity, public policy actions are governed by elaborate and often time-consuming processes and, depending upon the details of the proposed program, high levels of conflict may be generated. Therefore, if seismic rehabilitation is to be achieved through a formal policy adoption and implementation process, several additional issues must be addressed.

Typical Issues: Once it has been decided that a formal seismic rehabilitation program is necessary, a variety of political leadership, technical, process, enforcement, and equity issues must be faced in trying to forge a program that is both effective and acceptable. The questions typically revolve around the choice of a voluntary or mandatory approach, the standards to be followed, the length of time allowed for compliance (and penalties for noncompliance), the distribution of costs and availability of cost off-sets (subsidies, incentives, etc.), and the impacts of dislocation and business interruption.

How can proponents achieve a place for seismic re-habilitation on the often crowded political agendas of governing bodies and can they get favorable ac-tion? Issues compete for space on the agendas of key policy-makers and executives, be they corporate boards of directors and chief executive officers or public-sector elected or appointed bodies and admin-istrative managers. Leveraging a place for earth-quake safety, especially the subject of rehabilitating potentially hazardous buildings, is a key first step in what is usually a lengthy process. History provides suggestions on how to place seismic rehabilitation on decision-makers' agendas. Earthquakes, at least for a short time, open the well known "windows of oppor-tunity" by creating a change from the context of nor-mal operations. In the aftermath of an earthquake, all of the following heighten awareness, at least for a time: the experience of actual losses and concern about the vulnerability of other properties; the costs of repair, replacement, or relocation; paying the relief and recovery expenses; and the everyday experience of driving home through a disrupted community. In other words, disaster experience usually, but not al-ways, turns what earlier might have been abstract and uncertain notions of threat to concrete appreciations of risk and thereby opens that famous "window." Disaster experience alone, however, may not be suf-ficient; there have been notable earthquakes that have not resulted in significant actions to reduce future losses.

Sustained leadership clearly plays a major role in achieving seismic safety objectives. For example, as a youngster, Los Angeles City Council member Hal Bernson experienced the 1952 Arvin-Tehachapi earthquakes. Later he was shaken by the 1971 San Fernando earthquake. Representing a major portion of the San Fernando Valley, he adopted seismic safe-ty as an issue when he joined the city council, and he has provided sustained leadership ever since. Al-though it took a decade (1971-81), Bernson led the way to the enactment of the well known Los Angeles ordinance requiring the rehabilitation of URM bearing-wall buildings. More recently, Councilman

Bernson chaired the council's ad hoc Committee on Earthquake Recovery following the 1994 Northridge Earthquake. In the lead capacity, Bernson sponsored and shepherded through to adoption the ordinance requiring the rehabilitation of pre-1976 concrete tilt-up buildings (which were shown to have been a ma-jor problem as early as the 1971 earthquake).

Using an incremental approach to solve recognized problems has a long and well documented history in the United States. In fact, it is a common public pol-icy strategy often dictated by budgetary or other prac-tical realities. In the area of nonstructural seismic rehabilitation, there is a relatively recent (1994) ex-ample. With the goal of eventually broadening its application, the Silicon Valley Uniform Code Adop-tion Committee added a new section (3403.6) to the codes administered by all Santa Clara County build-ing departments. As a condition of tenant improve-ments, this new section states:

> When a permit is issued for alterations or repairs, the existing suspended ceiling system within the area of alteration or repair shall comply with the lateral design requirements of UBC Standard 25-2 Part III because this amendment is necessary to mitigate a known seismic hazard in existing build-ings.

At the state level in California, Senator Alfred E. Al-quist was a junior member of the Senate in 1969 when a staff member convinced him to adopt seismic safety as an issue, partly because no one else "had it" and partly because the staff member believed that earthquake safety had important statewide implica-tions. Alquist's efforts resulted in the 1970 creation of a powerless, token, legislative study committee, the Joint Committee on Seismic Safety. Nature, coincidence, or luck then took a hand. The February 1971 San Fernando earthquake suddenly highlighted the existence of this legislative study committee (which became immediately recognized and re-spected) and led directly to many of California's seis-mic safety policy changes. Included in the innova-tions and with then-Governor Ronald Reagan's con-currence was the "institutionalization" of seismic safety at the state level via creation of the California Seismic Safety Commission. The fundamental long-term change (bolstered by a series of damaging earth-quakes and widely publicized increasing probabilities) has been that seismic safety is now a legitimate and recurring item on the legislature's agenda.

Informal discussions suggest that this pattern of issue-adopting by key leaders exists in private-sector organizations as well. In some cases, the pressure to address the seismic rehabilitation of buildings (and other mitigation and preparedness activities) comes from the home offices of companies with facilities in active seismic areas.

Can local jurisdiction leaders adopt their own program or do they need authorizing legislation from a higher level? This fascinating intergovernmental relations issue is both real and symbolic. It may be that some states, partly because of their statewide building code requirements, would not permit local jurisdictions to adopt retroactive seismic rehabilitation ordinances without authorizing state legislation or without an initiative at the state level to empower local agencies to carry out such programs. In more decentralized states such as California, the cities of Los Angeles, Santa Rosa, and others have the power and took the initiative to enact rehabilitation requirements.

State action may either sanction a desired local initiative or, depending upon political context, provide an acceptable scapegoat for local officials, especially where policy action at the local level is hard to achieve. In the late 1970s, the California legislature, for example, enacted a law protecting design professionals and others involved in seismic rehabilitation from liability under specified conditions, and this facilitated an array of local actions by removing an inhibitor to the professional design community.

In many cases, local officials would prefer that the citizens perceive them as "having to carry out a state requirement" rather than as policy initiators themselves. At the same time, many state legislatures are dominated by suburban and rural members, and seismically hazardous buildings are not problems for their districts. Therefore, unless it is a very urban state, issues like the rehabilitation of buildings often do not receive full attention from state legislators, and it may be difficult to get state action. As one veteran of Utah's early seismic safety efforts noted, the Utah legislature primarily responds to local pressures rather than initiating much itself, especially if the members perceive an issue as infringing on "local

control." In this context, a strong consensus among local governments on the desired state action is critical. Again, the situation will determine how to approach the need for facilitating and/or authorizing legislation from higher levels.

Are there ways to accommodate the various interests in the process of program design? Seismically rehabilitating existing buildings, especially if they are occupied, can become complicated because of the temporary — and perhaps permanent — dislocations involved. In moving away from the private voluntary program, in which the owner controls the fate of the occupants, to the mandatory program, where the "we" versus "they" conflicting interests may become paramount, the rehabilitation process should be ready to deal with the range of issues and their advocates. While the specific situation will determine the cast of characters and their positions, they can range from employee groups who pressure for rehabilitation for their own protection (or oppose it because the relocation site may extend their home-to-work journeys) to low-income tenants of single-room occupancy (SRO) buildings whose mobility and options are very limited.

The heart of dealing with the range of potentially involved groups is to deliberately identify the various "stakeholder" interests in the rehabilitation process. A strategy then must be devised to include these group or their representatives, hear their concerns, and accommodate them to the extent possible in the project planning phase. Many local agencies, especially those involved with planning and community development, have extensive experience with citizen involvement and community hearings processes, and this experience can be tapped and adapted for proposed seismic rehabilitation projects.

It may be that some permanent dislocations will be necessary, and these will have to be evaluated on a project-by-project basis. Problems are lessened by the extent to which affordable and available (and often nearby) space is available, relocation assistance is provided, and the opportunity to return to the rehabilitated structure is "guaranteed" or at least offered to the previous occupants. Solving the "various interests" problem may require cooperative efforts between the building owners, real estate agents, property managers, and government officials.

What are the trade-offs between mandatory and voluntary programs? As noted above, this publication is intended to help the reader understand the basic choices available in seismic rehabilitation and the fact that as such projects move from the private voluntary model to the informal/encouragement model and, finally, to the fully mandated program model, levels of conflict and complexity increase. Nevertheless, each model has characteristic advantages and shortcomings. Even though greatly oversimplified, Figure 4 summarizes the "pros and cons" of each model.

Worthy of note is that this is not a linear sequence by any means. Owners may or may not choose to rehabilitate; local and state governments may or may not create formal programs (but they might lend encouragement and indirect support); local code and other administrators might establish threshold standards or criteria that are "triggered" on a case-by-case basis; and the federal government may seismically rehabilitate its buildings regardless of whether or not local jurisdictions do anything about seismic safety.

All rehabilitation costs money and it has to come from someone. The mandatory approach to rehabilitation is the most financially complex of the three largely because government becomes an increasingly important part of the solution and is therefore expected to bring its resources to the table. This expectation is especially high when the scope of seismic rehabilitation encompasses a relatively large number of buildings and prescribes potentially expensive rehabilitation standards.

Owner self-funding of seismic rehabilitation follows traditional paths and is of real concern only to the owner. Self-financing includes renegotiating the mortgage to generate rehabilitation funds, using current income or savings, borrowing on the commercial market, and/or selling additional stock to raise capital (if it is a stock company). Public financial assistance, however, comes in different forms and is constrained by laws and regulations that often prescribe in detail the allowable and legitimate purposes for which public monies may be expended. The underlying doctrine is that while governments can be partners in financing solutions to community problems, they cannot provide a gift of public funds for solely private ends. As is well known in public finance, capi-

tal facilities planning and the community development professions, the mixtures of government and private funding become very complicated. In actuality, the financial packages come to resemble—metaphorically—"marble cakes." As government's role increases in seismic rehabilitation so does that "marbelling." The challenge, therefore, is to define the respective roles of the private sector and government in seismic rehabilitation in ways that make it feasible for each to contribute to the goal of providing safer buildings in as affordable a manner as possible. There are both direct and indirect ways to do this, examples of which are discussed below.

In fully mandated programs, government's role as a partial financial partner can be critical. Local officials will have to consider the range of financial assistance they can offer to support the process. Oakland's seismic rehabilitation program for private buildings is stalled because no money is available to help owners with the costs. Meanwhile, the rehabilitation of Oakland's historic City Hall was financed partly by a combination of voter-approved local bond funds and federal disaster assistance monies which flowed from the 1989 Loma Prieta earthquake. San Francisco issued bonds, and San Jose has a redevelopment district in which URM building owners can get assistance in financing their engineering studies and rehabilitation projects.

Government officials have great experience in financing various projects. For example, direct methods include capital funding to provide new or upgraded facilities, issuing bonds to be repaid over several decades, securing matching funds from state and federal sources, and using tax increment financing. Indirectly, government can support the seismic rehabilitation process by working with lenders to create attractive loan programs for community purposes, waiving application and permit fees for projects, and providing transferable development credits. The essential point is that government financial managers and private sector companies must cooperate in seismic rehabilitation programs. In the long run, they could be each other's most important partners.

What are the incentives for compliance and penalties for noncompliance with a program? Incentives and penalties can take many and sometimes surprising forms, and the more formal the seismic rehabilitation

FIGURE 4 Seismic rehabilitation choices—advantages and limitations

VOLUNTARY PROGRAM	INFORMAL/ENCOURAGEMENT PROGRAM	MANDATORY PROGRAM
ADVANTAGES: • Clearly reflects policy that owners are ultimately responsible for the performance of their buildings. • Owner and design and construction team choose project scope, design criteria, timing, and process. • Limited governmental involvement or control over project, except for normal permitting requirements, but may trigger other requirements. • Owner assumes all project costs. • Process is comparatively simple and contains little conflict. • May help local economy and revitalization of the nearby area. • May set example for other owners. • Economic hardships not an issue. **LIMITATIONS:** • May reduce the risk, but not get desired level of earthquake resistance. • Independent technical review by building departments may be limited by lack of standards and expertise. • Few buildings are involved, and the pace of seismic rehabilitation can be slow. • Triggering of other requirements may kill the project.	**ADVANTAGES:** • Symbolizes a practical more flexible commitment than the mandatory approval. • Based on some form of seismic safety trigger (change of occupancy, percentage or remodeling, cost, etc.) • Owner assumes responsibility for project-related dislocations and relocations. • Provides for adherence to a set of common requirements that is based on some level of actual earthquake risk. • Allows variabilities of each building to be considered. • Provides for some level of independent design and construction review, assuming the expertise is available. • Few buildings make this relatively easy to administer on a case by case basis. • May be part of a local revitalization program that improves local economy. • While conflict may arise over a given project, widespread mobilization of opposing interests is avoided. • Costs borne by owners as part of total project costs or may be some sharing with government. • Completed projects could serve as examples for other owners considering extensive ("triggered") remodeling or rehabilitating projects. **LIMITATIONS:** • May reduce the risk, but not fully address actual risk. • Case by case approach may be slow and difficult to administer because each project is unique. • Local officials have no influence over potentially earthquake hazardous buildings unless they are going to be substantially remodeled. • May result in evictions and lease terminations, resulting in unforeseen community problems. • Requires fairly sophisticated expertise and assigned responsibilities in building departments. • Could involve involuntary dislocations and relocations with little due process available to those being displaced. • Does not represent a shared community commitment to seismic safety. • May change with rotation of building department personnel. • May result in owner relocating out of the jurisdiction to one where requirements do not exist.	**ADVANTAGES:** • Symbolizes a political (community-wide) commitment to seismic safety. • Government and owners may share costs, responsibility for project-related dislocations and relocations. • Is based on formal policy with specified standards and regulatory processes. • Each project is independently reviewed and inspected, assuming the expertise is available. • Results in lower earthquake losses and less demand for response and recovery services and money. • Assures uniformity of approach and adherence to a formal schedule for all parties resulting in a more predictable process. • May help revitalize local areas and economy. • May reduce the risk, but not fully address the actual risk. **LIMITATIONS:** • May create unrealistic earthquake performance expectations among the public and community leaders. • Is the most difficult to establish politically, and may be feasible only in high risk areas. • May involve direct or indirect cost sharing by local jurisdictions. • Depending on scope, can result in significant dislocations, which may be the local governments' responsibility to solve. • Rather than conform, some owners may abandon the properties, relocate to other jurisdictions without such requirements, or take other avoidance measures. • May result in evictions and lease terminations, resulting in unforeseen community problems. • Generates the highest level of conflict as the pool of affected interests is expanded. • Economic hardship can be very significant. • May result in higher rent and lease costs, making it even more difficult for lower income tenants and marginal businesses to survive. • May make it difficult for owners to sell, insure, or qualify for mortgages for nonrehabilitated properties. • While meeting the formal criteria, but by stimulating the seismic rehabilitation market, can result in questionably competent practitioners and projects. • May inhibit revitalization by adding costly requirements.

program, the more obvious are the incentives and penalties. However, even in the voluntary and encouraged approaches, important incentives/disincentives exist. The exact mixture depends, of course, upon the approach taken to seismic rehabilitation, but the content and roles of incentives and penalties should be carefully considered in the choice of program type and in the program design phase.

For example, publicizing voluntary rehabilitation may result in increased business and local goodwill (which may be used to achieve other purposes) or it might instill confidence in home office staff and suppliers and customers that a private facility will be capable of operating with a minimum of interruption after an earthquake. In another case, local government can create wealth indirectly by issuing "development credits" for multiple property owners who seismically rehabilitate their buildings. Indirect incentives also may include waiving other requirements (e.g., having to provide off street parking) or allowing the owners to add additional stories to a new building elsewhere. Government also can participate more directly in seismic rehabilitation by investing public funds in street lighting, transportation, landscaping, and other improvements as part of a broader areawide renewal effort; by establishing and guaranteeing discounted interest loan programs to help finance seismic rehabilitation; or by helping find suitable space and paying the direct costs of relocating businesses and residents from structures destined for seismic rehabilitation.

Penalties for not complying with required seismic rehabilitation requirements can be serious, but there is a general reluctance to use them except as a last resort. Most public policy in this specialized field relies on obtaining at least grudging building owner compliance by using realistic standards, providing practical time limits, offering independent appeals processes, and trying to find incentives and subsidies. Nevertheless, the range of potential penalties includes the nonissuance of permits until the plans address seismic rehabilitation requirements, condemnation and removal of the structure under the special provisions of "dangerous buildings" ordinances, issuance of court orders, and adding tax and other lien-type penalties to nonconforming properties. Interestingly, not all penalties have to be governmental. As conditions of a loan, some banks are requir-

ing risk analyses and earthquake insurance coverage that directly affect an owner's decision about buildings known to be earthquake-vulnerable.

How will the community benefit from seismic rehabilitation in the long run, and how can the short run dislocations of businesses and residents be ameliorated? The issue of long-term gain versus short-term pain pervades virtually all community renewal, revitalization, redevelopment, and restoration measures, not just seismic rehabilitation. The governmental process is the proper place to negotiate a balance between the short-term dislocations and longer-term benefits to the community. When seismic rehabilitation of buildings is made a component of larger processes or programs, it is much more likely to be successful.

Los Angeles, for example, paid close attention to the costs of its measures and established two increments of rehabilitation. The first step required — in a short time — the anchoring of the URM bearing walls to the floors and roof structures of the affected buildings, a comparatively inexpensive task that often could be accomplished without dislocating the occupants. The second step involved more extensive and expensive bracing and other measures but allowed installation over a longer time. Interestingly, the ordinance specified that owners who failed to meet the initial anchoring requirements had to meet the second set of requirements in less time than those who had complied, thereby providing a kind of incentive to move quickly on step one's basic anchoring.

Managing the Legal Issues of Seismic Rehabilitation

Nature of the Problem: The very nature of seismic rehabilitation focuses on modifying existing buildings — those built earlier and under different rules. Therein lay the potential legal problems that tend to cluster around the following:

• Potential liability,

• Building owners' rights to due process,

• Disclosure of known hazards,

• The taking of private property and unwarranted exercises of governmental police powers,

- Actions related to absentee landowners,

- The right of government to enact requirements above those sufficient to protect life,

- Gifting of public funds,

- Foreclosure proceedings,

- Negligence,

- Sovereign immunity,

- Foreseeability and unreasonableness of risk versus providing protection,

- Interpretations of "acts of God,"

- Discovery and statutes of repose,

- Causation and concurrent causation,

- Reasonableness of costs to carry out mandates, and

- Status of regulatory codes, design procedures, and similar materials and their use or enforcement as a standard of practice.

There are precedents for responding to a number of these issues, but the fundamental principle is to take only those actions that can be defended within existing state law or local ordinances. It is an axiom of America, however, that anyone has the right to sue anyone (despite some immunities); therefore, legal challenges to seismic rehabilitation should be expected.

Some working definitions are probably in order. In general, a "building code" is formally adopted legislation establishing standards and procedures that regulate the design, construction, alteration, and similar activities related to new and existing buildings. As such, codes are the "law of the land" in the adopting jurisdictions. "Guidelines," by contrast, serve multiple purposes, some of which may have legal implications. They provide users with peer-developed information about dealing with specific issues, in this case the seismic rehabilitation of existing buildings. In this capacity, guidelines serve to help educate users, provide them with a basis for taking appropriate actions, and serve as a common reference. To the extent that guidelines are widely and easily available, they

can be used to assess a design professional's knowledge of the state of the art in the field. Moreover, while the specific guidelines considered here, the *NEHRP Guidelines for the Seismic Rehabilitation of Buildings*, were not prepared to be a "model code," it would not be difficult for code-writing organizations and building officials to adapt them for such use. For example, the *Guidelines* would become a *de facto* code if a building official used them to accept or approve a proposed seismic rehabilitation project, especially if the proposer deviated from them without sound justification.

A "standard of practice" is more difficult to define because its use as its determination requires extensive judgment and information. In general, a standard of practice is a yardstick against which to measure or compare a practice or action. Everything else being equal, a user is expected in like circumstances to provide a standard of practice comparable to his/her peers.

However, throughout these legal discussions is the fundamental "reasonable person" principle. For example, judgments would be made on what a "reasonable person" would do or be expected to do under the following illustrative circumstances: the apparent probability that the harm-causing event will occur, whether the person involved actually knew or should have known the risk, the magnitude of the expected resulting harm, and the effort required to institute proper precautions.

Typical Issues: Legal challenges to seismic rehabilitation programs tend to revolve around several specific issues.

Can the local jurisdiction adopt and enforce regulations that require owners to rehabilitate their buildings when these very same buildings met whatever standards were in force at the time of their construction? This question goes to the heart of seismic rehabilitation as an issue of private cost versus public benefit. Moreover, in many cases, the state must be the adopting jurisdiction for any code.

Can the jurisdiction adopt building standards for existing buildings that are less stringent than those in force for new buildings? A positive answer implies a dual level of safety — people in newer buildings are safer than those in older buildings. While perfect safety is impossible to achieve, some types of older building perform better in earthquakes than others and, given the state of knowledge and practice of earthquake-resistant design, every earthquake teaches new lessons (witness the "steel frame buildings problem" after the 1994 Northridge earthquake). Ample justification can be adduced to require existing buildings to be strengthened for the common good. Comparable examples include requiring the retroactive installation of fire sprinkler systems, fire-resistant doors, and fire escapes.

What is the liability of design professionals and contractors performing seismic rehabilitation work that does not (and often cannot) meet the requirements of the current code in force for new buildings? Building codes sometimes contain triggers that may require a building to be brought up to current codes for new construction. Changes in materials, technology, design philosophy, construction methods, and a host of other factors may make it nearly impossible to both practically and economically upgrade a building to current standards. Historic buildings are even more of a challenge, but work on them is often governed by special codes and standards.

What happens if the rehabilitated building is damaged or causes death and injury in a future earthquake? This question anticipates that rehabilitation may prove at least partially ineffective, so great care must be taken to clarify the program objective as being to reduce -- not eliminate -- the potential loss of life and injury in an earthquake. Thus, if a rehabilitated building suffers less damage in an earthquake than it would have before being strengthened, even though it might be a total economic loss, it could be judged to have performed adequately. Moreover, the effectiveness of the rehabilitation most likely will be greater in smaller and perhaps more frequent earthquakes than in the very rare great event where the rehabilitated building could suffer serious damage but probably still less than it would have without any strengthening.

A study (*Life Safety and Economic and Liability Risks Associated with Strengthened Unreinforced Masonry Buildings*) completed in 1994 by the J. H. Wiggins Company is worth quoting in part for it provides particularly useful insights into real legal issues — at least in the California context — that arose following the 1989 Loma Prieta earthquake (pp. 124-130):

> Lawsuits that were filed in the aftermath of the Loma Prieta earthquake established that building owners and design professionals will be held accountable for damages and injuries as a result of structural failures during an earthquake. . . . The key to these large settlements was the fact that the owners could not rebut the abundance of notice they had concerning their buildings' structural defects and their failure to take remedial steps to mitigate the hazards presented by the buildings. . . After Loma Prieta, all UMB owners will be held liable for failing to take corrective measures to mitigate their buildings' hazardous condition. In addition, the owners' design professionals who have reviewed these buildings may be brought into lawsuits, both as defendants and percipient witnesses. . . . Litigation after the Loma Prieta earthquake demonstrated that jurors clearly understand that, under California law, codes are merely a minimum standard. Thus, actual jury reaction has demonstrated that mere code compliance will not be a sufficient defense to protect a property owner from liability. . . . Building owners who have delayed taking action to mitigate the hazards presented by their building's lack of seismic resistance may be faced with a claim of punitive damages if the building causes injuries in an earthquake. An injured occupant or passerby may contend that the owner had knowledge of his building's hazardous condition and was therefore guilty of willful and conscious disregard for the rights and safety of others. . . . To avoid claims of malpractice, design professionals must ensure that their work is done in accordance with the standards of the community in which they practice....Therefore, if a design professional such as an architect or engineer designs a retrofit (rehabilitation) plan using a lower level of safety (such as is contained in many local ordinances), the design professional could ultimately face a claim of liability for malpractice on the grounds that they employed a lower standard than that used in their community.

Solving the Problem: State laws and local ordinances plus precedent-setting decisions from elsewhere define how the legal issues related to seismic rehabilitation can be addressed in any given situation or locality. The key to minimizing legal problems and potentially lengthy delays in implementing seismic rehabilitation programs is to include legal counsel from the very outset.

Counsel will be heavily involved in preparing seismic rehabilitation ordinance language; explaining its provisions within the context of existing law; defending its principles and procedures throughout the policy formulation, adoption, and implementation phases of the seismic rehabilitation program; and answering any challenges that arise.

State and local governments can adopt ordinances and programs that require improvements to existing buildings for reasons of public safety. In general, the courts and legislatures understand that changes in technology, materials, and social needs (e.g., energy conservation and providing access for handicapped people) are legitimate public concerns and that building owners can be required under specified conditions to modify their structures accordingly.

The reality is that not everyone is equally safe. While it is important to narrow the gap, practical technical, political, and economic reasons can be offered for not requiring existing buildings to meet all of the requirements for new buildings. Clearly, the precedent has been set for state and local governments to adopt and enforce less-than-current-code requirements for existing buildings. *Uniform Code for Building Conservation* is a good example as are the court-tested seismic rehabilitation ordinances of Los Angeles and other communities. For a seismic rehabilitation program to be defensible, it must be demonstrated is that the requirements are for public benefit; are reasonable; are uniformly and fairly applied; and include provisions for exceptions, delays, or the use of equivalent alternative measures.

Design professionals and contractors worry a great deal about being held liable for the performance of buildings (and often pay high premiums for errors and omissions insurance). A concern of some design professionals is whether or not they are exposed to liability or criminal charges if a seismically rehabilitated building does not meet the current code's requirements for new construction. Most believe it is commendable to improve a building, and thereby increase safety even though they could not bring it up to the current code governing new construction. In general, however, the best defense is due diligence, adherence to requirements, a practical standard of care, and a test of reasonableness. These seem to be the issues around which most building-related controversies arise.

As noted earlier, partly to help remove this barrier, California enacted SB 445 which relieved local governments and design and construction personnel from liability when doing seismic rehabilitation work under less stringent standards than those required for new buildings. However, this immunity was not extended to cases where negligence or other unreasonable practices were found. Thus, while it is easy to provide general protection, the challenges will be on a case-by-case basis.

While earthquakes are natural events, it is human-designed and -built structures that cause the casualties and property losses. If losses are experienced in seismically rehabilitated buildings as they very well may be, it will be important to show that the project adhered to the requirements and that the work was properly performed. For example, seismically strengthened URM buildings in Los Angeles sustained damage in the Northridge earthquake and, even though the event fortuitously occurred early in the morning on a holiday, it is clear that in most cases the strengthening measures prevented more serious losses of life and injuries. In other words, they achieved the life-safety objectives of the program.

The bibliography in Chapter 6 includes some legal references directly related to seismic safety and building rehabilitation that will help the reader understand the general nature of the issues and determine when legal counsel should be consulted. The context of the particular policy decision or project will greatly determine the applicable legal issues and strategies for dealing with them.

CHOOSING THE TARGETS: SINGLE BUILDINGS, NEIGHBORHOODS, OR CLASSES OF BUILDINGS

Nature of the Problem: A strategic question that must always be answered when structuring a seismic rehabilitation program involves how narrow or broad will the scope be. The answer has significant implications for the policies and actions required, the standards to be applied, the availability of the skills needed, and other factors. Individual buildings can be dealt with on case-by-case basis, but prescribing seismic rehabilitation efforts for areas of town (e.g., Pioneer Square in Seattle), for specific types of building, (e.g., pre-1976 tilt-up wall structures in Los Angeles), or for specific occupancies (e.g., theaters or apartment buildings) is central to defining the rehabilitation program's objective, methods, and processes. The scope decision also will define the community interests that are affected by the decision (e.g., the local "apartment owners and managers association" if rehabilitating apartment buildings is to be the objective).

Typical Issues: Several issues should be considered in choosing the focus of a seismic rehabilitation program. In fact, one should expect that, for a variety of local reasons, the focus of the final seismic rehabilitation program may change during the program design and adoption phases. For example, early and powerful opposition from theater and apartment building owners and church leaders to an early version of the Los Angeles URM seismic rehabilitation ordinance (which attempted to focus on high-occupancy uses) actually caused proponents to broaden the scope to all URM buildings because the apartment, theater, and church representatives complained about being "singled out" unfairly. It also matters greatly if the program focuses solely on government buildings or affects the private sector as well.

In Salt Lake City, in addition to wanting to preserve the important and historic City and County Administration Building by renovating and seismically strengthening it (including a new seismic isolation foundation system), city officials hoped that the public project would provide an example to private owners of responsible actions taken on potentially hazardous buildings. The Church of the Latter Day Saints contributed to this process by voluntarily seis-

mically strengthening the former Hotel Utah, now used as a church office building. Questions that most likely will arise include the following.

Are we going to focus on classes or types of buildings, or specific uses or occupancies or on one or more geographic areas? While every building is unique, cities differ as well. The amply documented poor earthquake performance of URM structures combined with a post-1971 political opening in Los Angeles yielded the Division 88 seismic rehabilitation program focusing on that particular type of structure. Following the 1994 Northridge earthquake, the same approach was taken in the ordinance requiring that seismic improvements be made to early tilt-up concrete wall buildings (buildings whose poor performance had first been documented in the 1971 San Fernando earthquake). Since the Northridge event, the city of Los Angeles has been voluntarily strengthening several of its fire stations, providing an example of a use focus. Following its damaging 1969 earthquakes, Santa Rosa, California, partly because it already had a bounded redevelopment project area, city passed a local ordinance that required the evaluation and strengthening of several types of buildings in the older downtown area. Therefore, Santa Rosa adopted a program based on a geographic scope.

What is the inventory of the targeted buildings (e.g., what is the number of building potentially involved)? This is both a technical and strategic/political question. Collecting building inventory information can consume time and money. It may come as a surprise, but most building departments and other city agencies have not conducted a census of the community building stock. An exception was the city of Los Angeles, where officials were fortunate to have had a good census of its URM buildings because decades earlier the city had enacted an ordinance requiring the strengthening or removal of dangerous parapets and file information on each of the subject buildings was kept. Another exception was Santa Rosa, California, which had an accurate inventory of the buildings in the downtown redevelopment area because of the need to examine various occupancies and uses during the planning process.

Buildings can be structurally tricky and, at some point, the specific characteristics of a building must be determined before seismic rehabilitation plans can be prepared. Since the earthquake resistance of a building depends largely on its frame (which is hidden from view) and because drawings usually are not available (especially for old buildings), real analytical challenges ensue, but the *Guidelines* documents may be of some help in this respect. Facades and earlier renovations may further confuse the issue. Engineers often talk about being surprised — usually negatively — when they move from preliminary "windshield survey" data (to help establish an estimate of the number of buildings of a specific class) to conducting site-specific tests to collect information about particular buildings.

This issue relates directly back to the conflict model. Except for perhaps gaining voter approval for a bond issue to seismically rehabilitate some city building (e.g., fire stations in Salt Lake City or an historic city hall, in Oakland, California), the number of structures is important to understanding the size of the proposed program, the resources needed, and the interests that may be mobilized. It really matters if the scope is a few buildings out of perhaps thousands or 50 percent of a town's commercial downtown area, which was the case in Oroville, California, after its 1975 earthquake. In the Oroville case, the collection of inventory data was easy, but the mobilization of the opposition represented by the Oroville Property Owners Association which was composed of leading members of the town's commercial and political structure, effectively defeated any meaningful seismic rehabilitation program.

Are there any special characteristics of the structures such as designated historical buildings, high density, low-income housing or others? The individual complexity of communities must be accounted for in designing seismic rehabilitation programs. Special considerations must be given, for example, to those buildings that have been designated as historic, and an increasing complication is the designation of local "historic districts" (e.g., as San Diego's Gaslamp District or Claremont California's older commercial area) that often contain the area's oldest structures. In such cases, the ad-

vice of specialists in historic preservation is essential early in the definition of any large rehabilitation effort.

The issue of density and the economic characteristics of the residents and businesses are important factors. For example, because of its very high population density, large low income housing stock, cultural identity, political importance and numerous small shops, San Francisco's Chinatown, which consists of the city's many poorly constructed post-1906 earthquake URM buildings, poses an enormous socioeconomic challenge to seismic rehabilitation. On the other hand, the fashionable, upscale, high income, but still densely populated area of Georgetown in Washington, D.C., would pose different socioeconomic and political problems if seismic rehabilitation measures were proposed for that or similar areas.

What does local political experience indicate about which community interests will mobilize around which choice and how will their influence be felt? Throughout this discussion it has been mentioned in passing that seismic rehabilitation programs, which change the rules from when the affected buildings were first constructed, are capable of mobilizing various interests. These interest will vary from community to community, and the challenge is to anticipate which interests will mobilize, what initial positions they might take, and what can be done through incentives, compromise and a perceived fair due process to accommodate their concerns.

Public officials are well aware that hearings, town meetings, and other democratic mechanisms attract more opponents than supporters; therefore, one should not overlook the need to mobilize allies of seismic rehabilitation. Local geologists can help explain the threat, local engineers can help answer technical questions, local construction industry representatives can talk about jobs, local community groups of many different kinds can discuss the positive benefits of revitalization, and other local advocacy groups may be available to help balance the debate. In addition, the local media can be quite influential by thoroughly covering and supporting a proposed seismic rehabilitation program (e.g., *Los Angeles Times*), reporting but taking no position (e.g., *Oroville Mercury Register*), or paying virtually no attention to the issue (e.g., *Oakland Tribune* following the 1989 Loma Prieta earthquake). Note that "local" is used frequently in this context because there is a common tendency in public forums to discount visiting experts "who don't have to live here." Local champions are better when facing local opponents.

Will seismic rehabilitation be the primary focus or will it be an element of some broader community program (e.g., a comprehensive redevelopment program for a designated area)? There are examples of both strategies: Los Angeles simply moved on seismic rehabilitation of URM buildings; Santa Rosa added seismic rehabilitation to the upgrading requirements for its downtown redevelopment area; and the Clayton, Missouri, school district listed seismic rehabilitation as only one of the many reasons for asking the voters to support a bond issue. In the post-Northridge setting, Los Angeles' Community Redevelopment Agency (CRA) defined several project areas that will include seismic rehabilitation as one element of an overall improvement strategy for the designated areas. Consequently, readers are urged to give careful consideration to evaluating the alternative strategies available to achieve seismic rehabilitation.

OPTIMIZING MULTIHAZARD MITIGATION TO REDUCE RISK

Nature of the Problem: Mitigation is the prevention of future losses. While seismically rehabilitating buildings will help accomplish that goal for earthquakes, buildings also are exposed to such other hazards as river and coastal floods, hurricanes and high winds, fire, and tornadoes. Moreover, because the rehabilitation of existing buildings extends their lives, it increases the probabilities that the buildings will experience the effects of the other hazards. Whenever possible, therefore, it is in the national interest that rehabilitation include measures to better protect the structure from the multiple hazards to which it is exposed over its (rehabilitation-extended) lifetime. Note, however, that overall mitigation be-

comes complex when one mitigative action such as raising a building for flood protection purposes increases its exposure to earthquake damage if the work done is not properly designed to avoid both threats.

Typical Issues: Several questions should be addressed in a multihazard mitigation context when considering rehabilitation of a building for purposes of seismic protection:

To which other hazards is the site subject? This question is largely one of determining what hazards assessment information exists, where it is located, and whether the quality of the information is adequate for use in a specific rehabilitation project. For example, the City of Seattle negotiates the extent of rehabilitation of an existing building in which the goal is to achieve a balance of life-safety improvements. Along with seismic improvements — which may not be the most urgent need — could be those related to improved exiting, and fire resistance (e.g., the addition of fire sprinklers and alarms).

Are there any governmental, property insurance, or other requirements governing rehabilitation to mitigate future losses? This question can be answered only by checking with the governing (permitting) local jurisdiction or lending or insuring institutions about what, if any, requirements exist. The design team should not overlook the requirements of independently governed special districts such as flood control agencies, fire protection districts, and historic districts. State and federal requirements might exist, and the local jurisdictions often provide information about or referrals to other responsible agencies.

How can we ensure through the project planning and design phase that effective mitigation measures are addressed and that potential conflicts between various corrective measures are resolved? This becomes a key question for the design and construction team.

Are there any financial or other incentives to help achieve multihazard mitigation, and what are the benefits and costs of doing so? The answers to this two-part question relate directly to the cost of the rehabilitation project. On one hand, it needs to be determined if incentives, subsidies, or other measures exist to help offset the costs of hazard mitigation. On the other hand, benefit-cost analyses can be done to help determine if the mitigation of existing hazards will, given the probable exposure to future events, be a worthy investment.

Solving the Problem: A fundamental principle to observe in multihazard mitigation is to ensure that the project planning and design process addresses mitigation as part of the rehabilitation project. There may be requirements to do so (e.g., laws requiring the installation of fire sprinkler systems due to substantial changes in the use and occupancy of a building), but others may address hazard mitigation voluntarily as part of their decision to protect their investment, to increase market value, or to provide a rapid return to operations. A few specific suggestions are discussed below.

Obtaining information about the exposure of a given site or building to various hazards is critical to taking effective mitigation measures. Yet, the availability and quality of such information varies greatly from area to area, and it is very difficult to pull all the information from various sources together. For example, flood control agencies have maps showing potential inundation areas under various flood scenarios; city and county planning departments in California often have hazards information as part of their required "Safety Elements"; geography and engineering departments of colleges and universities have their own collections; consultants may have done studies for nearby sites; and state and federal agencies such as the Federal Emergency FEMA and the U.S. Army Corps of Engineers (COE) can be useful providers of hazards information. However, it is the project design team that will have to assimilate and synthesize this information to ensure that it is adequately addressed early in the rehabilitation project planning phase.

While floodplain regulations are the most widely known from a national perspective, many states and localities have specific site preparation and construction requirements designed to reduce the exposure to various threats. In addition, there are sufficient examples to show that property financing and insuring

organizations may require attention to hazard mitigation as a condition of their support. For example, a well-known western bank explicitly requires that environmental, asbestos, and earthquake hazards be assessed as a condition of a property loan. The key is to ensure that the question is thoroughly researched by the design team.

Mitigation efforts may disclose apparent conflicts between effective measures to deal with multiple hazards. Cutting holes in structural walls to add fire sprinkler systems may weaken the wall from an earthquake perspective or the pipes may break during an earthquake such as happened to an Oakland, California, building in the 1989 Loma Prieta earthquake because rigid fire sprinkler piping crossed through a seismic separation joint between two parts of what appeared to be, but was not, one building. Consequently, it is very important that the design team identify and resolve in the project planning stage potential conflicts between mitigation measures. This may require expert advice from practitioners in each field and their involvement from the very beginning of the process so that each understands the overall performance objectives and plans. They can then design their elements so as to minimize potential problems. Such coordination can virtually eliminate conflicts between mitigation actions taken for different purposes, especially now that the *Guidelines* documents are available for use in evaluating the seismic aspects of building safety.

Direct and indirect financial incentives may exist to promote multiple hazard mitigation. Their existence, however, is not universal and will have to be determined early in project planning. The small city of Torrance, California, for example, established an assessment district to help finance the seismic rehabilitation of older buildings within the district's boundaries. As noted earlier, state law in California excludes seismic improvements made to buildings from being reassessed for property tax purposes. These concepts could be expanded to include other types of safety-related rehabilitation. Other possibilities include bond funds, property exchanges, and benefits from redevelopment programs.

Chapter 5
APPLICATIONS SCENARIOS

Every seismic rehabilitation project occurs because someone has chosen or been required to modify a building. Because "every building has its own story," actual seismic rehabilitation projects depend upon the local societal and organizational contexts in which they take place. While the purpose of Chapter 3 was to present three alternative models to help the user of the *Guidelines* documents select a path through the forest of general issues related to seismic rehabilitation, this chapter narrows the focus and offers the reader a set of relevant scenarios that illustrate specific "typical" situations and highlight key factors important to achieving seismic rehabilitation. Although many variations are possible, these three scenarios (a private initiative, a local regulatory approach, and a professional service request) represent common seismic rehabilitation motivations and processes.

The first scenario focuses on a private voluntary decision. The facilities manager of a company owning 16 buildings in various cities across the United States received the *Guidelines* documents and wishes to determine if all or any of his buildings are possibly hazardous in earthquakes. If this proves to be the case, the facilities manager will recommend whether a seismic rehabilitation process be initiated with the company's own funds.

The second scenario addresses the public policy dilemma of a city manager whose chief building official received a copy of the *Guidelines* documents. After review and conference, they jointly decide to initiate the preparation of a proposed mandatory seismic rehabilitation ordinance for the city council's consideration.

The third scenario places a private consulting structural engineer, who knows little about earthquake engineering, in the difficult situation of needing to respond to his/her client by determining if any of the client's multiple properties in the Midwest is susceptible to earthquake damage. If so, the consulting structural engineer is to recommend whether any or all of the client's buildings should be seismically rehabilitated.

SCENARIO ONE: THE PRIVATE COMPANY

Situation

As the corporate facilities manager, you are responsible for all property acquisition, leasing, construction, remodeling, operations, and maintenance of the company's buildings. Your employer owns 16 buildings of various ages, sizes, and types of construction nationwide (Los Angeles, 5; Albuquerque, 1; Seattle, 2; St. Louis, 3; Charleston, 1; Baltimore, 2; and New Haven, 2).

Because of your position as facilities manager, you recently attended a workshop on seismic rehabilitation of existing buildings and you received the Guidelines documents. As a result, you became concerned about the potential earthquake performance of your company's buildings. The chief executive officer (CEO) has authorized you to evaluate the earthquake risk and likely earthquake performance of the 16 buildings. Your task is to assess the risk and likely earthquake performance of the 16 buildings and make seismic rehabilitation recommendations (which include doing nothing) to the CEO and possibly to the company's board of directors.

Considerations

Many factors have to be taken into account in your report which will influence the decision to invest or not invest in the seismic rehabilitation of the buildings. You may have to collect some information from other company units. Some of the issues you need to consider are:

- The geographic distribution of objective earthquake risk;

- The expected loads from the most likely seismic events;

- The probability of those events likely to occur (e.g., the planning horizon);

- The expected performance of the buildings from the expected earthquake loads;

- Competing needs for the funds and the trade-offs between short-term profits and long-term asset protection, including inventory and equipment values;

- The current status of capital replacement timetables and the flexibility of those timetables;

- Current business planning that could affect short-term and long-term use of the buildings (e.g., changes in product lines and markets, rates of facility obsolescence, and the existence or nonexistence of functional redundancy in other "safer" locations); and

- The benefits and costs associated with seismic rehabilitation.

You are aware that implementation of a voluntary seismic rehabilitation program within the company will require:

- Conducting a formal comparative risk evaluation and an initial screening or rapid assessment of the buildings;

- Developing an upgrading program that addresses various levels of desired performance;

- Specifying alternative design strategies to achieve those desired performance levels;

- Determining whether there are financial incentives external to the company that might be available for seismic rehabilitation;

- Determining what penalties external to the company may be imposed for not choosing to rehabilitate.

- Assessing the extent and depth of commitment to seismic rehabilitation of the company's top management and the board of directors; and

- Judging how and where seismic rehabilitation will fit in with and help meet the company's overall business objectives and priorities.

You are also aware that operational considerations must be factored into the decision about how to deal with the earthquake risk to the company's buildings by:

- Locating design professionals and contractors capable of performing seismic risk evaluations and the rehabilitation work;

- Determining if a seismic rehabilitation project will trigger requirements to comply *with other* local building code provisions that could add significantly to the costs and increase business interruption (e.g., disabled access, plumbing, electrical, life safety, asbestos removal, and energy conservation requirements);

- Estimating the costs of permits and inspections including the timeliness and difficulty of the process; and

- Assessing the value to the company of enhanced visibility and the goodwill associated with public knowledge that the company has engaged in a program of voluntary seismic rehabilitation of its buildings.

SCENARIO TWO: LOCAL GOVERNMENT POLICY DECISION

Situation

You are a city manager and generally aware that your community might experience periodic damaging earthquakes. Your chief building official has informed you that he has received and studied the recently issued *Guidelines* documents by the Federal Emergency Management Agency. The building official informs you that your community has two classes of exceptionally vulnerable buildings -- unreinforced masonry (URM) and early (pre-1973) concrete tilt-up light industrial buildings.

As the city's chief executive officer, you agree with the building official that an appropriate action would be to prepare an ordinance for city council consideration. The proposed ordinance would require the owners of these two identified classes of building to seismically rehabilitate them and to use the *Guidelines* to meet the ordinance's requirements. In effect, this course of action means that you and the building official have to prepare the proposed ordinance; serve as the city's lead staff members for advising the council on the technical, socioeconomic, and other issues likely to arise if the ordinance is passed; and be ultimately responsible for enforcement of the "Community Earthquake Rehabilitation Ordinance."

As city manager, your experience tells you that *regardless* of the merits of a proposed ordinance to require the strengthening of URM and early tilt-up buildings, enacting and implementing it will be highly controversial. You also know that for the ordinance to both pass and then be effectively implemented, the city will need political leaders and a coalition of supporters behind the proposal.

Considerations

You and the building official have to be prepared to explain to the city council, media, and the public several important items:

- The earthquake threat to the community;

- What other communities facing a comparable threat are doing about the problem;

- The community-wide benefits of avoiding future losses, the costs of doing nothing, and the costs of rehabilitation;

- Plans to address the unique problem of historic buildings;

- The capabilities of local design professionals and contractors to meet the provisions of the ordinance;

- Ways to ameliorate the dislocations and economic effects caused by rehabilitation; and

- The need for rapid improvement of your staff's technical abilities.

From a program implementation perspective, you will have to address several other points including:

- The minimum level of compliance;

- The square foot costs and how costs will be shared, if at all, by building owners and the city;

- What other upgrade requirements will be triggered;

- The capabilities of city staff and whether staff will need to be increased and how;

- The appeal and arbitration procedures;

- The length of time for compliance;

- For what period of time owners will be exempt from additional retroactive measures; and

- The process and cost for handling noncomplying buildings (e.g., through condemnation and demolition).

Interestingly, this scenario demonstrates why jurisdictions often use "nonmandatory" alternatives to achieve the goal of seismic rehabilitation. For instance, an ordinance might only require that owners of buildings in the two suspect classes have licensed architects or structural engineers evaluate the buildings and file with the city reports that then become a matter of public record. This strategy could result in the quasivoluntary strengthening of buildings because the owners possess "guilty knowledge" of the susceptibility of their buildings, knowledge that could raise questions of liability associated with an existing hazard should a damaging earthquake occur.

SCENARIO THREE: THE CONSULTING ENGINEER'S DILEMMA

Situation

You are a consulting engineer in a small midwestern town located in a low seismic zone. Because of your professional interests, however, you are aware of specialist peers in the field of "earthquake engineering." Moreover, you are aware that the New Madrid fault zone, which has received a lot of publicity of late, is about 200 miles away.

While a particular concern for earthquakes has not been part of your lengthy practice, one of your best long-term clients has raised the earthquake issue with you. Following the client's attendance at a seminar on New Madrid area earthquakes at the University of Memphis' Center for Earthquake Research and Information where she obtained a copy of the newly released *Guidelines* documents, your client is concerned about the earthquake resistance of her apartment and commercial buildings located in Memphis, St. Louis, Kansas City, and several other smaller cities in the same general area. The client is concerned about the area's earthquake risk and her responsibilities and liabilities as a property owner.

Considerations

This situation is a real dilemma for both you as the consulting engineer and your client. Some of your key considerations include:

1. Getting more exact risk information;

2. Defining other skills needed to augment your own and their availability;

3. Determining if the cities where the buildings are located require seismic rehabilitation and if so, to what level;

4. Determining whether other code requirements will be triggered by work undertaken to seismically strengthen the buildings; and

5. Determining, now that you are a "knowing person," what, if any, liabilities are associated with the earthquake performance of your client's buildings.

Further considerations relate to evaluating client's properties; establishing priorities based on risk, occupancy, function, and other factors; determining acceptable levels of performance under expected events; designing effective rehabilitation schemes; accurately estimating costs; determining whether seismic rehabilitation can somehow be linked to the owner's general long-term property improvement plans; and deciding whether advising your client to sell the properties is a viable solution. Clients seldom understand that there are no guarantees in earthquake engineering and especially in the seismic rehabilitation of existing buildings. The consulting engineer who oversees a seismic rehabilitation project always has lingering concern about what will happen when an earthquake does occur and a rehabilitated building does not perform to the client's expectations. For example, a California Seismic Safety Commission report (p. 49) noted that "many engineers view the performance of retrofitted buildings in the Northridge earthquake positively" but "many owners were unaware that a retrofitted (rehabilitated) building could still be damaged to the point of not being economically repairable." One way to lessen this concern is for the design professional and the client to understand that, just as with the performance of new buildings, the effectiveness of seismic rehabilitation will vary with the severity of the earthquake. To illustrate this point, FEMA's benefit-cost volumes note that the anticipated effectiveness of an investment in seismic rehabilitation varies with the intensity of an earthquake. The greatest economic benefit derives from rehabilitation measures that perform best in lower magnitude but more frequent events. For example, rehabilitating a common low-rise tilt-up building is expected to reduce damages by 50 percent at modified Mercalli intensity (MMI) VI but only 30 percent at MMI XII.

Chapter 6
SELECTED ANNOTATED BIBLIOGRAPHY AND ADDITIONAL REFERENCES

The various "societal" (political, socioeconomic, administrative, and policy) problems inherent in the seismic rehabilitation of buildings and discussed in this publication are treated in literature that can be considered a subset of the literature on earthquake hazard mitigation which, in turn, is a subset of the literature on natural hazard mitigation. Thus, in discussing seismic rehabilitation or "hazardous structure abatement," there are three distinct but partially overlapping sets of reference literature that, taken together, are quite extensive.

The purpose of this publication has been to alert and orient the reader and potential user of the *Guidelines* documents with the array of societal problems often encountered in the seismic rehabilitation of buildings. A full treatment of each component of the array, however, simply is not feasible in a single document.

Once an individual begins to address seismic rehabilitation, he/she will face many of the problems and issues discussed earlier in this volume. The first section of this chapter presents a selected annotated bibliography designed to help those individuals identify appropriate additional reading, most of which also contain reference lists. It focuses on a core group of 10 books, 4 chapters from another book, 13 journal articles, and 4 reports. The second section of this chapter presents a list of other excellent works that may be of use to readers in specific situations.

CORE READINGS

A place to start exploring the policy and socioeconomic issues involved in the seismic rehabilitation of buildings is a January 1996 Earthquake Engineering Research Institute publication, *Public Policy and Building Safety*, an excellent and very readable report that succinctly surveys all of the major technical (i.e., nonengineering) issues and suggests practical strategies for understanding and dealing with many of

them. It includes a case study of the development of the Los Angeles ordinance requiring the inspection of steel-frame buildings; an overview of the typical policy-making process; and a reminder-style checklist of social, economic, and political factors to be considered in building safety.

An unusual and intentionally thought-provoking 1989 essay by Timothy Beatley, "Towards a Moral Philosophy of Natural Disaster Mitigation," appears in the *International Journal of Mass Emergencies and Disasters* (7 March 1989: 5-32). It is a clear and well written exploration of a rarely asked but fundamental question: What is the extent of government's moral obligation to protect people and property from natural disasters such as hurricanes and earthquakes? While many of the examples are drawn from the hurricane milieu (Beatley's specialty), mitigating the earthquake risk is addressed as well. Beatley argues that mitigation as public policy may be built on four ethical bases: utilitarian and market failure rationales (maximizing net social benefits); the concept of basic rights (providing primary physical security and subsistence); culpability and the prevention of harm (highlighting responsibility and costs); and paternalism (legitimating government interventions).

A more conventional starting place is with a book by William J. Petak and Arthur A. Atkisson, *Natural Hazard Risk Assessment and Public Policy: Anticipating the Unexpected* (New York, New York: Springer-Verlag, 1982), which describes and explains mitigation policies and programs within the larger context of disasters and/or disaster management.

A "handbook" spelling out a four-step mitigation process (community analysis, emergency analysis, mitigation needs assessment, and mitigation strategy development) is, *Practical Mitigation: Strategies for Managing Disaster Prevention and Reduction* (Rockville, Maryland: Research Alternatives Inc.,

1982) by James W. Morentz, Hugh C. Russell, and Judith A. Kelly. The orientation of this work is much more practical than conceptual. Of special interest are 81 mitigation case histories from across the United States involving all types of natural and technological hazards.

A special kind of "cookbook" (meant in the best possible sense) cosponsored by FEMA and the International City Management Association is *Emergency Management: Principles and Practice for Local Government* (Washington, D.C.: International City Management Association, 1991) edited by Thomas E. Drabek and Gerald J. Hoetmer. This comprehensive textbook is intended for "front line" emergency managers and local government officials. The "Introduction" and Part I, "History and Foundations of Emergency Management," provide the reader with basic concepts and terminology setting the stage for the remaining parts. Two chapters are of specific relevance to mitigation -- Chapter 5, "Perspectives and Roles of the State and Federal Governments," which explains in detail the relationship between local emergency management and other levels of government, and emphasizes the intergovernmental process and system interdependence, and Chapter 6, "Disaster Mitigation and Hazard Management," which covers the evolution of federal mitigation policy, the relationship between mitigation and comprehensive emergency management, hazard identification and analysis, and mitigation strategies, tools and techniques.

In *Natural Hazards and Public Choice: The State and Local Politics of Hazard Mitigation* (New York, New York: Academic Press, 1982), Peter H. Rossi, James D. Wright, and Eleanor Weber-Burdin explore attitudes of "political influentials" toward hazard mitigation across 20 states using 100 community samples and 2,000 respondents. Their findings that community elites across the United States did not find hazards to be very important compared to other problems and that these elites preferred "quick fixes" to politically painful long-term measures were subsequently challenged by Elliott Mittler in *Natural Hazard Policy Setting: Identifying Supporters and Opponents of Nonstructural Hazard Mitigation* (Boulder, Colorado: University of Colorado Institute of Behavioral Science, 1989). Using the same data but a more sophisticated statistical treatment, Mittler

came to different, more positive conclusions about elite hazard perceptions and about who tends to be a supporter, nonsupporter, or neutral with respect to hazard policy.

A special edition (Volume 45, January 1985) of a leading scholarly journal, *Public Administration Review*, is entitled "Emergency Management: A Challenge for Public Administration," with William J. Petak serving as editor. This issue is an excellent overview/primer devoted to FEMA and to disaster response and recovery (including technological disasters). Of the 21 articles, all relatively short, those dealing at least in part with mitigation are: "Emergency Management: A Challenge for Public Administration" by William J. Petak; "Emergency Management and the Intergovernmental System" by Alvin H. Mushkatel and Louis F. Weschler; "Disaster Recovery and Hazard Mitigation: Bridging the Intergovernmental Gap" by Claire R. Rubin and Daniel G. Barbee; "Mitigation Strategies and Integrated Emergency Management" by David R. Godschalk and David J. Brower; and "Financing Disaster Mitigation, Preparedness, Response, and Recovery" by Allen K. Settle.

Continuing a focus on intergovernmental issues and problems is a thoughtful 1984 article by William J. Petak, "Natural Hazard Mitigation: Professionalization of the Policy Making Process," in *International Journal of Mass Emergencies and Disasters* (2, August: 285-302). In this article Petak examines the constraints/barriers to adopting and implementing hazard mitigation policies. Petak notes that while FEMA historically has pushed state and local governments to improve mitigation and enhance response and recovery capabilities in order to better handle hazards on their own, those very same state and local governments are constrained by geophysical, ecological, and sociopolitical factors. With this in mind, Petak addresses two important questions: How can current and projected natural hazard losses be reduced through improvements in building and land use practices in designated hazard areas? How can the adoption and use of specific hazard mitigation approaches by state and local governments be accomplished?

Also treating the intergovernmental problems generated by disaster is *Disaster Policy Implementation:*

Managing Programs Under Shared Governance (New York, New York: Plenum Press, 1986) by Peter J. May and Walter Williams. Adopting a "two worlds of disaster politics" approach (the world of normal politics/low saliency and the world of active policy making in the aftermath of a disaster), this study was driven by two fundamental questions: How are good ideas turned (or not) into concrete actions? How might FEMA stimulate greater mitigation and preparedness efforts? Taking an "implementation perspective," May and Williams explore the "politically less visible aspects of disaster policy" under situations of "shared governance" (local, state, and federal).

Perhaps the core book of the 1980s is Thomas E. Drabek's *Human System Responses to Disaster: An Inventory of Sociological Findings* (New York, New York: Springer-Verlag, 1986). This work is a self-conscious attempt to survey the disaster literature extant at the time and create an "encyclopedia" of findings. It remains a fundamental resource in the field, and significant attention is focused on to mitigation.

Next is a book edited by Louise K. Comfort, *Managing Disaster: Strategies and Policy Perspectives* (Durham, North Carolina: Duke University Press, 1988). This collection of original essays by 21 scholars in the field of public policy is organized around two basic questions: What are the primary issues confronting public managers in a disaster? What actions/measures can they take to save lives and protect property? Case studies are woven into the articles, and significant attention is paid to mitigation.

W. Henry Lambright began a research project in the early 1980s on the rapidly evolving role of states (including California) in disaster management, and he subsequently published *The Role of States in Earthquake and Natural Hazard Innovation at the Local Level: A Decision-Making Study* (Syracuse, New York: Syracuse Research Corporation, 1984, also available from the U. S. Department of Commerce, National Technical Information Service). Lambright's logic of comparison is actually based on three different "policy settings": Emergent (South Carolina and Nevada); intermediate (California); and advanced (Japan). The core of the study is the application of a six-stage process of innovation model em-phasizing "entrepreneurs," "triggers," "the search for options," "adoption," "implementation," and "incorporation."

Focusing solely on one California policy innovation, Lambright followed his larger study with a 1985 journal article, "The Southern California Earthquake Preparedness Project: Evolution of an 'Earthquake Entrepreneur'" in the *International Journal of Mass Emergencies and Disasters* (3, November: 75-94). Lambright depicts the Southern California Earthquake Preparedness Project as a novel mechanism created to accelerate the pace and intensity of preparedness.

Kathleen J. Tierney reviews much of the mitigation literature through 1989 in "Improving Theory and Research on Hazard Mitigation: Political Economy and Organizational Perspectives "in the *International Journal of Mass Emergencies and Disasters* (7, November 1989: 367-396). In this article, Tierney notes that mitigation is the least studied and therefore the least understood of the four key disaster phases. The literature on mitigation, according to Tierney, can be divided into three major areas: studies on public perceptions of mitigation measures; research on agenda setting, adoption, and the implementation of hazard mitigation measures; and studies assessing the impact of hazard mitigation measures. Moreover, three themes pervade the literature on disaster mitigation: the only slightly coupled relationship between perceived risk and level of mitigation; the difficulty in promoting mitigation programs because the problems they attempt to address are complex and highly technical; and the positive role played by critical events in the adoption and implementation of hazard mitigation programs.

Questioning the role of critical events is Elliott Mitler in *The Public Policy Response To Hurricane Hugo In South Carolina* (Boulder, Colorado: University of Colorado Institute of Behavioral Science, Natural Hazards Research and Applications Information Center, Working Paper 84, April 1993). This study contradicts the popular assumption that in the honeymoon period following a major disaster, political windows open easily for mitigation improvements. He maintains that those windows do not always open and, even if they do open, they slam shut very quickly.

Another antidote (but from earthquakes and from California no less) to the facile assumption that disasters lead easily to mitigation improvements is *Standing Rubble: The 1975-1976 Oroville, California Experience with Earthquake-Damaged Buildings* (Sacramento, California: Robert Olson Associates, Inc., 1988) by Robert A. Olson and Richard Stuart Olson. An article-length version appeared as "The Rubble's Standing Up in Oroville, California: The Politics of Building Safety" by Richard Stuart Olson and Robert A. Olson in the *International Journal of Mass Emergencies and Disasters* (11, August 1993: 163-188).

Another book high on any "must read" list for earthquake mitigation is *Earthquake Mitigation Policy: The Experience of Two States* (Boulder, Colorado: University of Colorado Institute of Behavioral Science, 1983) by Thomas E. Drabek, Alvin H. Mushkatel, and Thomas J. Kilijanek. This book is important not only because it pays explicit attention to definitions and policy issues, but also because its selection of state cases does not include California. In fact, hitting head-on the tendency to think of earthquake mitigation and California as synonyms, the authors subtitled their Missouri chapter, "This Isn't California," and their Washington chapter, "North from California." Rich in detail, the authors discuss three case histories of conflicts over earthquake mitigation policy that reveals the perceptual barriers and resource constraints typical at the state and local levels. Of particular interest is Chapter V, "Resistance from Below: St. Louis vs. HUD," which chronicles an intergovernmental political battle over lateral force requirements for building rehabilitations.

Almost a decade later, Philip R. Berke and Timothy Beatley published *Planning for Earthquakes: Risk, Politics, and Policy* (Baltimore, Maryland: Johns Hopkins University Press, 1992). Combining micro and macro approaches, Berke and Beatley present three earthquake mitigation case studies (Salt Lake County, Utah; Palo Alto, California; and Charleston, South Carolina) with statistical analysis of the responses to a questionnaire on mitigation practices from 202 communities in 20 states.

Arnold J. Meltsner's, "The Communication of Scientific Information to the Wider Public: The Case of Seismology in California," in *Minerva* (3, Autumn

1979: 331-354) follows the early 20th century history of seismology studies in California and the tremendous political obstacles faced by earth scientists and engineers who attempted to convince California's leaders to publicly recognize and come effectively to grips with the earthquake threat. The article chronicles the truly heroic efforts to establish that most basic of earthquake mitigation policies -- a seismic building code -- and is an excellent antidote to the myth that California's road to seismic safety prominence was easy.

The issue of what to do about "bad buildings" constitutes a small but important literature of its own. Still the only book-length study of the policy dilemmas inherent in trying to reduce the life-safety threat posed by unreinforced masonry buildings is *The Politics and Economics of Earthquake Hazard Mitigation: Unreinforced Masonry Buildings in Southern California* (Boulder, Colorado: University of Colorado Institute of Behavioral Sciences, Monograph 43, 1986) by Daniel J. Alesch and William J. Petak. In this book, Alesch and Petak analyze three California cases: Long Beach, Los Angeles, and Santa Ana. The emphasis is on the interplay between technical solutions, the economics and financing of building rehabilitation, and the political maneuvering (especially the role and importance of the "window opening" San Fernando earthquake of 1971) that yielded different ordinance outcomes in each of the cities.

To be read as a companion piece to Alesch and Petak's book is Richard Stuart Olson's, "The Political Economy of Life Safety: The City of Los Angeles and 'Hazardous Structure Abatement,' 1973-1981" in *Policy Studies Review* (4, May 1985: 670-679). Taking a more explicitly political viewpoint than Alesch and Petak, Olson profiles the "pro" and "con" sides on the famous Los Angeles seismic rehabilitation ordinance and emphasizes the importance of a credible scenario for a future earthquake to the passage of the Los Angeles ordinance.

The last item in the core list is the February 1994 "theme issue" of *Earthquake Spectra*. Edited by Mary C. Comerio, this journal issue reflects the outcome of a U.S.-Italy workshop held in October 1992 and focuses on "Design in Retrofit and Repair." The contributions revolve around 10 problems that both U.S. and Italian experts had to confront: achieving a balance between life safety and cost, achieving a balance between life safety and building conservation, developing strategies "to preserve existing buildings (not just monuments)," finding support for pre-design investigations by an entire design team in preparation for formatting rehabilitation designs, developing performance criteria for building systems and for historic preservation as complements to structural design criteria, insufficient understanding of materials performance, insufficient understanding of the performance of composite structures resulting from multiple retrofits, resolving incongruities between finite elements analysis and building failure typologies, insufficient understanding of building performance over multiple earthquakes and how better information on that issue should be incorporated into reconstruction codes, and determining whether the building will be lost in another earthquake or by the engineer's design?

ADDITIONAL READINGS

Natural Hazards

Unique in the field and almost falling in the core list (except that it is 660 pages) is James Huffman's *Government Liability and Disaster Mitigation: A Comparative Study* (Lanham, Maryland: University Press of America, 1986). Undertaken by a professor of law, this is a fascinating study of liability laws and how they affect assignment of costs and, therefore, mitigation policy in six countries -- New Zealand, the United States, Peru, Japan, China, and what was then the Soviet Union.

In 1985, Peter J. May published *Recovering From Catastrophes: Federal Disaster Relief Policies* and Politics (Westport, Connecticut: Greenwood Press, 1985). In this work May asks who wins and who loses when it comes to bearing the costs and risks of disaster relief. Tracing the political evolution of disaster relief policy, May examines three histories -- legislative, organizational, and, most interesting, "what really happened." The legislative history focuses on policy changes, congressional politics, and the driving question of the federal government's appropriate role in disaster relief.

Another general treatment of the disaster problem in the United States is Raymond J. Burby's, *Sharing*

Environmental Risks: How to Control Governments' Losses in Natural Disasters (Boulder, Colorado: Westview Press, 1991). Summarizing the results of an extensive study of the losses from over 130 natural disasters occurring in the 1980s, Burby analyzes the complex relationship between federal, state, and local policies. While the work is comprehensive, Part II, "How to Control Losses," is dedicated to mitigation and focuses on the problem of how "to ease the perennial hardships states and localities suffer." A short chapter, "The Special Case of Earthquakes," argues that earthquakes create consequences and problems different from those caused by floods, hurricanes, and landslides. The author then addresses how earthquake-prone local governments can be persuaded to insure their property at risk.

Earthquake Hazard Mitigation

Also almost falling in the core list is a recent book by Robert A. Stallings, *Promoting Risk: Constructing the Earthquake Threat* (New York, New York: Aldine de Gruyter, 1995). Starting from a different base than the other authors, Stallings explores why earthquake risk has not achieved the status of a fully developed "social problem" given the likely national consequences of a catastrophic earthquake. For Stallings, the answer is that "promoters" of the earthquake threat have followed essentially an "insider" strategy and not a "grass-roots" strategy and have therefore failed to generate widespread public support.

Another study notable for its non-California intent is Arthur A. Atkisson and William J. Petak's "The Politics of Community Seismic Safety" in *Proceedings of Conference XV: Preparing for and Responding to a Damaging Earthquake in the Eastern United States* (Reston, Virginia: U.S. Geological Survey, Open-File Report 82-220, 1982).

Other specific but non-California studies include those by Peter J. May and others in, *Earthquake Risk Reduction Profiles: Local Policies and Practices in the Puget Sound and Portland Areas* (Seattle, Washington: University of Washington, Institute for Public Policy and Management, November 1989) and *Anticipating Earthquakes: Risk Reduction Policies and Practices in the Puget Sound and Portland Areas* (Seattle, Washington: University of Washington,

Institute for Public Policy and Management, November 1989).

Also worth reading is a short article by Peter J. May and Patricia Bolton, "Reassessing Earthquake Reduction Measures," in the *Journal of the American Planning Association* (52 Autumn 1986: 443-451), and May's "Addressing Public Risks: Federal Earthquake Policy Design" in the *Journal of Policy Analysis and Management* (10, Spring 1991: 263-285).

A basic resource document on federal efforts to promote seismic safety, that contains much original information is, *To Save Lives And Protect Property: A Policy Assessment of Federal Earthquake Activities, 1964-1987* (Washington, D.C.: Federal Emergency Management Agency, 1988) by Robert A. Olson, Constance Holland, H. Crane Miller, W. Henry Lambright, Henry J. Lagorio, and Carl R. Treseder.

Two U. S. Geological Survey studies that emphasize knowledge transfer and applications are *Applications of Knowledge Produced in the National Earthquake Hazards Reduction Program: 1977-1987* (Reston, Virginia: U.S. Geological Survey Open File Report 88-13-B, 1988) edited by Walter W. Hays and *Applications of Research from the U.S. Geological Survey Program, Assessment of Regional Earthquake Hazards and Risk Along the Wasatch Front, Utah* (Reston, Virginia: U.S. Geological Survey Professional Paper 1519, 1993) edited by Paula Gori. For further reading on the surprisingly partisan politics of seismic safety in Utah, see Richard Stuart Olson and Robert A. Olson's,

"Trapped in Politics: The Life, Death, and Afterlife of the Utah Seismic Safety Advisory Council" in the *International Journal of Mass Emergencies* (12, March 1994: 77-94).

A significant comparative work is *Earthquake Mitigation Programs in California, Utah, and Washington* prepared by C. E. Orians and Patricia A. Bolton for the Workshop on Issues and Options for Earthquake Loss Reduction (Seattle, Washington: Battelle Human Affairs Research Center, BHARC-800/92/041, September 1992).

In the same vein is a study by Joanne M. Nigg and others, *Evaluation of the Dissemination and Utilization of the NEHRP Recommended Provisions* (Washington, D.C.: Federal Emergency Management Agency, May 1992).

Agency reports to the U S. Congress often are given short shrift as resources, but some are of high quality. Such is the case of a 1993 FEMA report, *Improving Earthquake Mitigation, A Report to Congress* (Washington, D.C.: FEMA, Office of Earthquake and Natural Hazards, January 1993). Noteworthy within that report are "Social Science Research: Relevance for Policy and Practice" by Russell Dyness, "Local Public Capacity to Deal with a Catastrophic Earthquake" by Claire Rubin and "Education, Awareness and Information Transfer Issues" by Paula Schultz.

Of historic interest are two federal reports from the 1970s. Stimulated by unexpectedly high losses in the 1971 San Fernando earthquake, the federal government began to pay more systematic attention to the earthquake problem in the United States. *Earthquake Prediction and Public Policy* (Washington, D.C.: National Academy of Sciences, 1975) was prepared by National Research Council, Panel of the Public Policy Implications of Earthquake Prediction of the Advisory Committee on Emergency Planning and *Earthquake Hazards Reduction: Issues for an Implementation Plan* (Washington, DC: 1978) was prepared in response to the *National Earthquake Hazards Reduction Act of 1977* (PL 94-125) by the Executive Office of the President, Office of Science and Technology Policy, Working Group on Earthquake Hazards Reduction.

California Studies

Thirty-one years before the Loma Prieta earthquake captured the world's attention, Karl V. Steinbrugge published *Earthquake Hazard in the San Francisco Bay Area: A Continuing Problem in Public Policy* (Berkeley, California: Institute of Governmental Studies, University of California, 1968).

An interesting California mitigation (land use) case study is presented by Martha L. Blair and William E. Spangle in *Seismic Safety and Land-Use Planning, Selected Examples From California* (Reston, Virginia: U.S. Geological Survey Professional Paper 941-B, 1979).

In 1980, as a result of the devastation wrought by Mount St. Helens earlier that year, President Carter turned even more federal attention to the earthquake threat in California. As a result, FEMA produced a slim but important document, *An Assessment of the Consequences and Preparations for a Catastrophic California Earthquake: Findings and Actions Taken* (Washington, D.C.: FEMA, November 1980). The essence of this report is a set of earthquake scenarios with associated probabilities and with estimated casualty (dead and injured) figures.

In 1983, the small central California town of Coalinga was virtually destroyed by an earthquake. The response was unusually draconian -- level it and start over. Kathleen J. Tierney chronicles the impacts and aftermath in *Report on the Coalinga Earthquake of May 2, 1983* (Sacramento, Califorina: California Seismic Safety Commission, 1985).

Multiple jurisdiction/intrastate studies of response to risk are rare, but two were authored in the mid-1980s: "Earthquakes and Public Policy Implementation in California," by Alan J. Wyner in the *International Journal of Mass Emergencies and Disasters* (2 August 1984: 267-284) and *Preparing for California's Earthquakes: Local Government and Seismic Safety* (Berkeley, California: University of California Institute of Governmental Studies, 1986) by Alan J. Wyner and Dean E. Mann.

Although most of the world will forever associate the 1989 earthquake in northern California with the baseball World Series, coincidentally between San Francisco and Oakland, that event is technically called the Loma Prieta earthquake. In the aftermath, Patricia A. Bolton and C. E. Orians undertook a study of that disaster's mitigation lessons: *Earthquake Mitigation in the Bay Area: Lessons from the Loma Prieta Earthquake* (Seattle, Washington: Battelle Human Affairs Research Center, Summary Report BHARC-800/92/015, March 1992).

On the same disaster but with a narrower focus on housing, Mary C. Comerio published "Hazards Mitigation and Housing Recovery: Watsonville and San Francisco One Year Later," in *Disasters and the Small Dwelling* (London: James and James Science Publishers, 1992) edited by Yasemin Aysan and Ian Davis.

As Executive Director of the California Seismic Safety Commission at the time, L. Thomas Tobin also reflected on the lessons of the 1989 disaster in "Legacy of the Loma Prieta Earthquake: Challenges to Other Communities," *Symposium on Practical Lessons from the Loma Prieta Earthquake* (Oakland, Califorina: Earthquake Engineering Research Institute, March 1993).

Also stimulated by the Loma Prieta event and ensuing lessons was *Use of Earthquake Hazards Information: Assessment of Practice in the San Francisco Bay Region* (Portola Valley, California: Spangle Associates, July 1993) by Spangle Associates.

The relationship between earthquake disasters and mitigation opportunities inherent in reconstruction is the theme of two other reports by Spangle Associates: *PEPPER: Pre-Earthquake Planning for Post-Earthquake Rebuilding* (Sacramento, California: California Office of Emergency Services, for the Southern California Earthquake Preparedness Project, 1987 and *Rebuilding after Earthquakes, Lessons from Planners* (Portola Valley, California: Spangle Associates, 1991).

As part of its own planning efforts, the California Seismic Safety Commission published and made widely available its *California at Risk, Reducing Earthquake Hazards 1992 to 1996* (Sacramento, California: California Seismic Safety Commission, Report SSC 91-091, 1992). From the same source and interesting from an historical viewpoint is *Earthquake Hazards Management: An Action Plan for California* (Sacramento, California: California Seismic Safety Commission, September 1982). Probably of the greatest historical import, however, is the California Legislature Joint Committee on Seismic Safety's *Meeting The Earthquake Challenge* (Sacramento, California: Legislature, State of California, January 1974). This study, commissioned as a result of the 1971 San Fernando earthquake, was really the blueprint for seismic safety improvements in California for more than a decade.

No list of literature on California would be complete or credible if it did not include *Waiting for Disaster: Earthquake Watch in California* (Berkeley, California: University of California Press, 1986) by Ralph H. Turner, Joanne M. Nigg, and Denise Heller Paz. This book addresses the issue of seismic preparedness in the high risk zone of Palmdale, California. Due to the alternating uplifting and subsiding of the earth's crust in the region (the so-called Palmdale Bulge), it was widely believed that Palmdale was a harbinger of earthquakes. Hypothesizing that this "near prediction" heightened the saliency of the region's earthquake threat, the authors examine the attitudes and actions of people and organizations in response to the threat.

Hazardous Buildings Studies

For more general reading on the conflict potential inherent in public policy attempts to deal with existing earthquake-vulnerable buildings, see Richard Stuart Olson and Douglas C. Nilson's "California's Hazardous Structure Problem: A Political Perspective," in *California Geology* (April 1983: 89-91), and subsequently reprinted in *Building Standards* (52, July-August 1983: 15-17).

How the federal government approached and handled the problem of its own earthquake-vulnerable buildings is the subject of Diana Todd and Ugo Morelli in "Adoption of Seismic Standards for Federal Buildings: Issues and Implications" in *Proceedings, Fifth U.S. National Conference on Earthquake Engineering, 1994* (Oakland, Califorina: Earthquake Engineering Research Institute, 1994, pp. 995-1003). In the same *Proceedings* (pp. 1005-1012) is another paper with a non-California focus -- David O. Knuttunen's, "New Code Provisions for Existing Buildings in Massachusetts."

Dealing with the problem of seismic rehabilitation of hospitals in an even more non-California (i.e., a non-United States) setting is Allan Lavell's, "Opening a Policy Window: The Costa Rican Hospital Retrofit and Seismic Insurance Program 1986-1992" in *The International Journal of Mass Emergencies and Disasters* (12, March 1994: 95-115). This article is especially interesting for its treatment of Costa Rica's ability to "learn" not only from its own earthquakes, but also from the Mexico City disaster of 1985.

Reflecting on housing lessons from the Los Angeles hazardous structure abatement ordinance is Mary C. Comerio in "Impacts of the Los Angeles Retrofit Ordinance on Residential Buildings" in *Earthquake Spectra* (8, February 1992: 79-94). In the February

1994 *Earthquake Spectra* theme issue discussed above in the core list, Comerio followed upon this earlier work with "Design Lessons in Residential Rehabilitation (pp. 43-64), which focuses on mitigation policy and housing in the aftermath of the 1989 Loma Prieta earthquake.

Example Rehabilitation Ordinances and Initiatives

To illustrate the array of subjects discussed in this publication, numerous enacted or proposed laws and ordinances and accompanying materials, bond issue descriptions, public finance materials, environmental impact reports, special studies, and federal documents and reports have been examined. While too voluminous to actually reprint in this *Societal Issues* volume, each is summarized below to make it as easy as possible for readers to understand the contents of these materials and to obtain any that might be of help.

City of Los Angeles, *Los Angeles Building Code, Chapter 88: Earthquake Hazard Reduction in Existing Buildings*, is available from the Department of Building and Safety, Building Bureau, 200 N. Spring St., Los Angeles, California 90012, (310) 485-2304. This well-known ordinance, enacted in 1981 (10 years after San Fernando earthquake), established a comprehensive program to require the seismic rehabilitation or demolition of unreinforced masonry bearing wall buildings built before 1934 (or for which a building permit was issued prior to October 6, 1933). The intent is clear: Where the analysis determines deficiencies, this chapter of the building code requires the building to be strengthened or demolished. The ordinance sets minimum standards, provides procedures and standards for identifying and classifying subject buildings according to their current use, provides analysis methods and allowable values, specifies information to be included on plans, defines priorities and time periods for compliance, and specifies penalties for noncompliance.

City of Los Angeles, Los Angeles Building Code, Division 91: *Earthquake Hazard Reduction in Existing Tilt-Up Concrete Wall Buildings* available for the Department of Building and Safety, Building Bureau, 200 N. Spring St., Los Angeles, CA 90012, (310)

485-2304. Similar in concept to Chapter 88, this ordinance focuses on another proven earthquake vulnerable building -- the tilt-up concrete wall buildings "designed under building codes in effect prior to January 1, 1976." The intent to require strengthening or demolition is the same. Like Chapter 88, Division 91 sets minimum standards for identifying and classifying subject buildings according to current use, provides analysis methods and allowable values, specifies notification procedures, prescribes information to be included on plans, defines priorities and times for compliance, and specifies penalties for noncompliance.

City of Los Angeles, Los Angeles Building Code, Proposed (June 16, 1994) *Chapter 92: Prescriptive Provisions for Seismic Strengthening of Light, Wood-Frame, Residential Buildings* available from the Department of Building and Safety, Building Bureau, 200 N. Spring St., Los Angeles, California 90012 (310) 485-2304. This ordinance, proposed following the Northridge earthquake, was adopted August 27, 1996, as a voluntary ordinance. It focuses on particularly vulnerable older light wood frame buildings that have the following structural weaknesses: "(a) sill plates or floor framing which are supported directly on the ground without an approved foundation system. (b) a perimeter foundation system which is constructed of wood posts supported on isolated pad footings. (c) perimeter foundation systems that are not continuous." Damage often is serious to structures with any of these characteristics, and the displaced occupants will result in a major demand for emergency shelter. This is a voluntary program, but like the city's other ordinances, this one also specifies analytical procedures and similar matters. Being prescriptive in nature the ordinance specifies how the corrective work should be done. Even though not officially adopted, it has been used as a handout and as a reference during plan checking.

City of Palo Alto, California *Ordinance Number 3666 adding Chapter 16.42 to the Palo Alto Municipal Code Setting Forth a Seismic Hazards Identification Program*, is available from the Building Inspection Division, 250 Hamilton, Palo Alto, California 94303, (415) 329-2550. While not able to enact a mandatory seismic rehabilitation program, Palo Alto succeeded in requiring that engineering reports be done and publicly filed by owners of the following

three types of buildings: all URM buildings, all pre-1935 buildings with 300 occupants or more other than URM buildings with 100 occupants or more, and all buildings constructed between January 1, 1935, and August 1976. The 1986 ordinance, anchored in the intent of the safety element of the city's comprehensive plan, defines responsibilities, scope, building categories, reporting requirements, review processes, and other matters.

City of Oakland, California *Ordinance Number 11274, Adopting Interim Standards for the Voluntary Seismic Upgrade of Existing Structures*, is available from the City Clerk, One City Hall Plaza, Oakland, California 94612(510) 238-3611. Ordinance 11274 was enacted in 1990 after the 1989 Loma Prieta earthquake. It was part of a series of policy efforts to deal with damaged buildings and to initiate a comprehensive program to abate the hazards posed by URM structures. This ordinance provides standards and force levels for upgrading, defines historic buildings to be exempted, establishes a design review and appeals process, and contains an exemption from future seismic upgrades for 15 years. It was seen as an interim measure until a permanent program could be established. One of the ordinance's goals was to "promote public health, safety and welfare," but this was to be done "within the constraint of reasonable economic effects."

City of Oakland, California Ordinance 11613, Adding Article 6 to Chapter 18 of the Oakland *Municipal Code Adopting a Seismic Hazards Mitigation Program for Unreinforced Masonry Structures* available from the City Clerk, One City Hall Plaza, Oakland, California 94612 (510) 238-3611. Ordinance 11613 is the city's URM building ordinance. It applies to all such buildings built before November 26, 1948 (the date of the city's first code containing seismic provisions), interestingly addresses both voluntary (limited scope) and mandatory (broader scope) rehabilitation standards, assigns interpretive responsibility to the building official, specifies right of entry, establishes notification and reporting requirements, establishes a public list of subject buildings and criteria for deletion of the building, establishes procedures for reviewing historic buildings, and provides for a variety of appeals and other processes.

State of California, *Health and Safety Code, Chapter 12.2 - Building Earthquake Safety* ("The URM

Law"), in available from legal research services or the California Seismic Safety Commission, 1900 K Street, Suite 100, Sacramento, California 95814, (916) 322-4917. Added to California's statutes in 1986, this law requires the building departments in all cities and counties located wholly or partially in the *Uniform Building Code Seismic* Zone 4 to "(a) identify all potentially hazardous buildings within their respective jurisdiction on or before January 1, 1990, (b) establish a mitigation program for potentially hazardous buildings to include notification to the legal owner, . . . and (c) by January 1, 1990, all information regarding potentially hazardous buildings and all hazardous building mitigation programs shall be reported to the appropriate legislative body of a city or county and filed with the Seismic Safety Commission." It requires the commission to monitor the program by annually publishing a report and was amended in 1993 to require that, upon transfer of ownership of any URM built before January 1, 1975, the purchaser must be given a copy of the *Commercial Property Owner's Guide to Earthquake Safety*. The law also refers to the following one, which excuses locals from associated liabilities.

State of California, *Health and Safety Code, Article 4 (Sections 19160 through 19168) - Earthquake Hazardous Building Reconstruction*, is available from legal research services or the Seismic Safety Commission, 1900 K Street, Suite 100, Sacramento, California 95814, (916) 322-4917. This law was passed in 1979 and was one of the earliest attempts to remove barriers to seismic rehabilitation. It was permissive in that the statute authorizes (not mandates) local jurisdictions to assess their hazards, allows for adoption of rehabilitation standards less than those required for new buildings, and among other subjects provides immunity from liability for local jurisdictions arising from damages to rehabilitated buildings or casualties caused by earthquakes. While well intended, the law also became an excuse for many local jurisdictions to do nothing until stronger legislation was passed in 1986.

U.S. Government, Office of the President, Executive Order 12941, *Seismic Safety of Existing Federally Owned or Leased Buildings*, is available from the Mitigation Directorate, Federal Emergency Management Agency, 500 C Street, S.W., Washington, D.C. 20472, (202) 642-3231. Based on earlier legislation,

this Presidential Executive Order is an example of the exercise of authority that could be provided to any chief executive, administrative officer, city manager, or other appropriate official. Executive Order 12941 sets minimum standards for use by federal departments and agencies "in assessing the seismic safety of their owned or leased buildings and mitigating unacceptable risks. . . " In addition, the order assigns implementation responsibilities, provides for periodically revising the standards, and requires the preparation of cost estimates consistent with the standards.

State of California, *Health and Safety Code, amending Section 18938 and adding Articles 8 and 9 to Chapter 1 of Division 12.5 Relating to the Rehabilitation, Changed Use, or Closure of Acute Care General Hospitals by January 1, 2030*, is available from legislative reference services or the Office of Statewide Health Planning and Development, 1600 Ninth Street, Sacramento, California 95814, (916) 654-3362. Following the 1971 San Fernando earthquake, state legislation was passed effective January 1, 1973, requiring new hospitals to be designed, reviewed, and constructed to higher standards. Later known as the "Alfred E. Alquist Hospital Seismic Safety Act," these amendments were passed in 1994 following the Northridge earthquake. By far, the most significant feature is the law's retroactivity: ". . . after January 1, 2008, general acute care hospital buildings that are determined to pose certain risks shall only be used for nonacute care hospital purposes" and ". . . no later than January 1, 2030, owners of all acute care inpatient hospitals shall demolish, replace, or change to nonhospital use, all hospital buildings that are not in substantial compliance, or seismically retrofit them so that they are in compliance with the [Office's] standards."

State of California, State Government Code, Sections 8878.50-8878.107, *Earthquake Safety and Public Buildings Bond Act of 1990 (Proposition 122)*, is available from the California Seismic Safety Commission, 1900 K Street, Suite 100, Sacramento, California 95814, (916) 322-4917. Added to California's statutes directly by its voters, this $250 million bond issue's purposes were to: "fund retrofitting, reconstruction, repair, replacement, or relocation of state-owned buildings or facilities which have earthquake or other safety deficiencies" and "provide financial assistance to local governments for earthquake safety improvements in structures housing those agencies critical to the delivery of essential government functions in the event of emergencies or disasters." The statute also funds related research and specifies how priorities, eligibility, fund distribution, and accountability will be maintained.

School District of Clayton, Missouri *Bond Issue Proposals*, available from the District's Community Relations Department, 75 Maryland Ave., St. Louis, Missouri 63105, (314) 726-5210. Of potential use to jurisdictions interested in seismic rehabilitation, but in lower seismic zones, this $18,365,000 bond issue "built in" earthquake resistance improvements to schools as part of a broader agenda. The agenda encompassed the need to accommodate increasing enrollment, to comply with the Americans with Disabilities Act (ADA), to preserve and properly maintain existing schools, to provide student access to modern computer technology, and "the obligation to protect lives of students in the event of an earthquake by strengthening portions of existing schools which do not conform to current building codes."

City and County of San Francisco, Department of City Planning, *Earthquake Hazard Reduction in Unreinforced Masonry Buildings: Program Alternatives*, Final Environmental Impact Report 89.112E, available from the City Planning Department, 1660 Mission St., San Francisco, California 94103, (415) 558-6287. This extremely valuable assessment of the community impacts of a proposed ordinance to require at least partial seismic rehabilitation of URM buildings contains a wealth of information on the issues discussed generally in this publication. One section, "Existing Financing Sources for the Retrofit of San Francisco's Unreinforced Masonry Buildings," was very helpful.

City of Oakland, California, Office of Public Works, *Preliminary List of Financial Resources to Consider in Developing a Local URM Seismic Safety Program*, available from the Office of Public Works, One City Hall Plaza, Oakland, California 94612, (510) 238-3961. Similar to the section of San Francisco's EIR, this list of potential funding alternatives and sources was prepared for the city by the staff of the California Seismic Safety Commission. It contains many of the same references as San Francisco's but also has additional information and some discus-

sion of the purposes and advantages and disadvantages of various financing mechanisms.

Federal Emergency Management Agency, *A Benefit-Cost Model for the Seismic Rehabilitation of Buildings, Volume 1, A User's Manual and Volume 2, Supporting Documentation* (FEMA 227 and 228), is available from the Publication Distribution Facility, 500 C St. S.W., Washington, D.C. 20472, (800) 480-2520. Increasing use is being made of methods to evaluate the benefits and costs of investing public funds, in this case for the seismic rehabilitation of private buildings. Later publications (FEMA 255 and 256) expand the use benefit-cost methods to federally owned buildings. These volumes provide background information and procedures and software for calculating the benefits and costs of seismic rehabilitation. The second volume in each set provides additional supporting data and technical papers.

Further References

In addition to the key items in the preceding annotated bibliography, there exists a myriad of other valuable materials used in preparing this publication. These included the following:

Association of Bay Area Governments (ABAG). 1979. *Attorney's Guide to Earthquake Liability*. Oakland, California: ABAG.

Association of Bay Area Governments. 1988. *Liability of Local Governments for Earthquake Hazards and Losses: Background Research Reports*. Oakland, California: ABAG.

Association of Bay Area Governments. 1989. *Liability of Local Governments for Earthquake Hazards and Losses: A Guide to the Law and Its Impacts in the States of California, Alaska, Utah, and Washington*. Oakland, California: ABAG

Association of Bay Area Governments. Seismic Safety Commission. 1992. *The Right to Know: Disclosure of Seismic Hazards in Buildings*, SSC 92-03. Sacramento, California: SSC.

Association of Bay Area Governments. 1994. *Incremental Seismic Retrofit Opportunities*. Silver Spring, Maryland: Building Technology, Inc.

Association of Bay Area Governments. February 1994. "Design Lessons in Residential Rehabilitation," *Earthquake Spectra* 10:1:43-64 (Earthquake Engineering Research Institute, Oakland, California).

Association of Bay Area Governments. 1995. *Northridge Housing Losses: A Study for the California Governor's Office of Emergency Services*. Berkeley: University of California.

Association of Bay Area Governments, Seismic Safety Commission. 1995. *Status of the Unreinforced Masonry Building Law: 1995 Annual Report to the Legislature*, SSC 95-05. Sacramento, California: SSC.

Beatley, Timothy, and Philip Berke. 1990. "Seismic Safety Through Public Incentives: The Palo Alto Seismic Hazard Identification Program," *Earthquake Spectra* 6:1: 57-79 (Earthquake Engineering Research Institute, Oakland, California).

Building Systems Development, Inc. 1993. *Reconstruction After Earthquakes: Issues, Urban Design, and Case Studies*. San Mateo, California: Building Systems Development, Inc.

Building Technology, Inc. 1994. *Facilities Management of Existing School Buildings: Two Models*. Silver Spring, Maryland: Building Technology, Inc.

Center for Environmental Design Research. 1987. *Unreinforced Masonry Seismic Strengthening Workshop and Cost Analysis*. Berkeley: University of California.

City and County of San Francisco, Department of City Planning. 1991. *Earthquake Hazard Reduction in Unreinforced Masonry Buildings: Program Alternatives, Final Environmental Impact*, Report 89.112E. San Francisco, California: San Francisco Department of City Planning.

City of Los Angeles, Housing Department. 1995. *Rebuilding Communities: Recovering from the Northridge Earthquake, January 17, 1994*. Los Angeles, California: City of Los Angeles Housing Department.

Coalition to Preserve Historic Long Beach. 1987. Achieving Urban Conservation and Minimizing Earthquake Risk (seminar materials). Long Beach, California: Coalition of Preserve Historic Long Beach.

Comerio, Mary C. 1989. *Seismic Costs and Policy Implications*. Oakland, California: George Miers and Associates.

County of Los Angeles, Chief Administrative Office. 1995. *Northridge Earthquake: Report of Findings, Joint Cities and County Hazard Mitigation Workshop*. Los Angeles, California: County of Los Angeles.

Davenport, Clifton W., and Theodore C. Smith. 1985. "Geologic Hazards, Negligence, and Real Estate Sales," *California Geology*, July (Department of Conservation, Division of Mines and Geology, Sacramento, California).

Dean, William E. 1995. "Marketable Risk Permits for Natural Disaster Mitigation," *Natural Hazards* 11:2: 193-201 (Kluwer Academic Publishers, The Netherlands).

Earthquake Engineering Research Institute. 1995. *Outline of Papers: 4th Japan/United States Workshop on Earthquake Hazard Reduction*. Oakland, California: EERI. See especially "The Path From Technical Knowledge to Implementation of Hazard Reduction Measures" by Catherine Bauman, "Post-Earthquake Sheltering and Housing: Will Some Special Population Groups Have Special Needs" by Patricia A Bolton, "Factors Affecting the Scope of Seismic Rehabilitation Projects and Programs" by Craig D. Comartin, "The Role of Local Government in Recovery of Housing" by Charles D. Eadie, "Standardization of Post-Earthquake Management Measures" by Laurence M. Kornfield, and "Immediate Response and Search and Rescue" by Richard A. Ranous.

Federal Emergency Management Agency. 1992. A *Benefit-Cost Model for the Seismic Rehabilitation of Buildings*, two volumes, FEMA 227 and 228. Washington, D.C.: FEMA.

Federal Emergency Management Agency. 1992. *Building for the Future*, Washington, D.C.: FEMA

Federal Emergency Management Agency. 1993. *Improving Earthquake Mitigation: Report to Congress*, Washington, D.C.: FEMA.

Federal Emergency Management Agency. 1992. *NEHRP Handbook for Seismic Rehabilitation of Existing Buildings*, FEMA 172. Washington, D.C.: FEMA.

Federal Emergency Management Agency. 1992. *Seismic Rehabilitation of Buildings -- Phase I: Issues Identification and Resolution*, FEMA 237. Washington, D.C.: FEMA.

Federal Emergency Management Agency. 1992. *State Participation in the National Earthquake Hazards Reduction Program*, Washington, D.C.: FEMA.

Federal Emergency Management Agency. 1994. *Seismic Rehabilitation of Federal Buildings: A Benefit/Cost Model*, two volumes: FEMA 255 and 256. Washington, D.C.: FEMA.

Federal Emergency Management Agency. 1995. *Benefit-Cost Analysis of Hazard Mitigation Projects*, Volume 5, Earthquake, User's Guide Version 1.01. Washington, D.C.: FEMA.

Federal Emergency Management Agency. 1995. *Typical Costs for Seismic Rehabilitation of Buildings, Second Edition*, FEMA 156 and 157, Washington, D.C.: FEMA.

Flatt, William D., and Wesley R. Kliner. 1991. "When the Earth Moves and Buildings Tumble, Who Will Pay? -- Tort Liability and Defenses for Earthquake Damage within the New Madrid Fault Zone." *Law Review*, 22:(1):1-41. (Memphis State University, Memphis, Tennessee).

H. J. Degenkolb Associates, Engineers. 1992. "Balancing Historic Preservation and Seismic Safety." *Details*. H. J. Degenkolb Associates, San Francisco, California.

Hart, Gary C., and Rami M. Elhassan. 1994. "Seismic Strengthening with Minimum Occupant Disruption Using Performance Based Seismic Design Criteria," *Earthquake Spectra* 10:(1):65-80 (Earthquake Engineering Research Institute, Oakland, California).

Hoover, Cynthia A. 1992. *Seismic Retrofit Policies: An Evaluation of Local Practices in Zone 4 and Their Application to Zone 3*. Oakland, California: EERI.

J. H. Wiggins Company. 1994. *Life Safety and Economic and Liability Risks Associated with Strengthened Unreinforced Masonry Buildings*. Redondo Beach, California: J. H. Wiggins Company.

Look, David, ed. 1991. *The Seismic Retrofit of Historic Buildings Conference Workbook*. San Francisco, California: Western Chapter of the Association for Preservation Technology.

Low, Ball and Lynch. 1989. *An Earthquake Seminar: What You Need to Know to Respond to Claims*. San Francisco, California: Low, Ball and Lynch.

Olshansky, Robert B., and Paul Hanely. Nd. "Seismic Building Codes," *Reducing Earthquake Hazards in the Central U.S.* Urbana-Champaign, Illinois: department of Urban and Regional Planning, University of Illinois.

Olshansky, Robert B. 1994. "Seismic Hazard Mitigation in the Central United States: The Role of the States." *Investigations of the New Madrid Seismic Zone*, U.S. Geological Survey Professional Paper 1538-F-G. Reston, Virginia.: USGS.

Olson, Robert A.. 1987. *Hazardous Buildings: No Longer a Technical Issue*. Sacramento, California: Robert Olson Associates, Inc.

Olson, Richard S., and Robert A. Olson. 1996. "What To Do?" *Politics, Public Policy, and Disaster: Oakland, California and the 1989 Loma Prieta Earthquake*. Tempe: Arizona State University.

Rutherford and Chekene Consulting Engineers. 1990. *Seismic Retrofitting Alternatives for San Francisco's Unreinforced Masonry Buildings: Estimates of Construction Cost and Seismic Damage*. San Francisco, California: Rutherford and Chekene.

Schulman, Irwin. Nd. *Engineering Liabilities for Design of a Bridge That Failed During the San Fernando Earthquake*. Los Angeles: State of California, Department of Transportation.

Scott, Stanley, and Robert A. Olson, Eds. 1993. *California's Earthquake Safety Policy: A Twentieth Anniversary Retrospective, 1969-89*. Berkeley: Earthquake Engineering Research Center, University of California.

Sharpe, Roland L. Nd. Seismic *Strengthening of Palo Alto Civic Center*. Cupertino, California: Engineering Decision Analysis Company, Inc.

Siegel, Henry I., and Lawrence W. Fowler. 1994. "Jackson Brewery: Upgrading an Historic Unreinforced Brick Building," *Earthquake Spectra* 10:(1):81-92, (Earthquake Engineering Research Institute, Oakland, California).

Solomon, K. A., and D. Okrent. 1977. *Catastrophic Events Leading to De Facto Limits on Liability.* Los Angeles: University of California, School of Engineering and Applied Science.

State of California, Office of Emergency Services. 1992. *Seismic Retrofit Incentive Programs: A Handbook for Local Governments.* Oakland, California: ABAG.

State of California, Seismic Safety Commission. 1992. *Architectural Practice and Earthquake Hazards: A Report of the Committee on the Architect's Role in Earthquake Hazard Mitigation.* Sacramento, California: SSC.

State of California, Seismic Safety Commission. 1990. *Earthquake Hazard Identification and Voluntary Mitigation: Palo Alto's City Ordinance*, SSC 90-05. Sacramento, California: SSC.

State of California, Seismic Safety Commission. 1987. *Financial and Social Impacts of Unreinforced Masonry Building Rehabilitation, SSC 87-02.* Sacramento, California: SSC.

State of California, Seismic Safety Commission. 1987. *Guidebook to Identify and Mitigate Seismic Hazards in Buildings and Appendix,* SSC 87-03. Sacramento, California: SSC.

State of California, Seismic Safety Commission. 1991. *Policy on Acceptable Levels of Earthquake Risk in State Buildings.* Sacramento, California: SSC.

State of California, Seismic Safety Commission. 1994. *Provisional Commentary for Seismic Retrofit*, SSC 94-02. Sacramento, California: SSC.

State of California, Seismic Safety Commission. 1985. *Rehabilitating Hazardous Masonry Buildings: A Draft Model Ordinance*, SSC 85-06. Sacramento, California: SSC.

Tierney, Kathleen J. 1993. *Socio-Economic Aspects of Hazard Mitigation.* Newark: Department of Sociology and Disaster Research Center, University of Delaware.

U.S. Department of Commerce, National Institute of Standards and Technology. 1995. *ICSSC Guidance on Implementing Executive Order 12941 on Seismic Safety of Existing Federally Owned or Leased Buildings,* ICSSC RP 5, NISTIR 5734. Gaithersburg, Maryland: NIST.

U.S. Department of Commerce, National Institute of Standards and Technology. 1994. *Standards of Seismic Safety for Existing Federally Owned or Leased Buildings and Commentary,* ICSSC RP 4, NISTIR 5382. Gaithersburg, Maryland: NIST.

U.S. Department of the Interior, National Park Service. 1994. *The Secretary of the Interior's Standards for Rehabilitation and Guidelines for Rehabilitating Historic Buildings.* Washington, D.C.: NPS.

U.S. Department of the Treasury. 1995. "Administration Policy Paper: Natural Disaster Insurance and Related Issues", joint letter from James L. Witt, Director, Federal Emergency Management Agency, and Robert E. Rubin, Secretary, Department of the Treasury, February 16.

VSP Associates, Inc. 1994. State and Local Efforts to Reduce Earthquake Losses: Snapshots of Policies, Programs, and Funding, A Report for the Office of Technology Assessment, U.S. Congress. Sacramento, California: VSP.

William Spangle and Associates, Inc. 1990. Strengthening Unreinforced Masonry Buildings in Los Angeles: Land Use and Occupancy Impacts of the L.A. Seismic Ordinance. Portola Valley, California: William Spangle and Associates, Inc.

Appendix A
THE FOUR STEPS IN DETAIL

THE FOUR STEPS

Step 1: Define the Problem

Step 1A: Preliminary Analysis

The measures outlined below are recommended as a starting point. The initial assumptions, estimates, and information collected may be informal, but as the endeavor proceeds to subsequent steps, the information should be improved.

Determine the probability of damaging earthquakes and determine whether it is significant enough to justify further action.

Request a formal statement on seismic risk from the U.S. Geological Survey (USGS), a state geological agency, a university professor of seismology, or a consulting seismologist or risk analyst.

Locate a map that depicts the location of faults and the intensity of ground shaking associated with an earthquake. The USGS, a state geological survey, FEMA, and other organizations have these maps or can help locate them.

Establish criteria, types of buildings considered to be unacceptably vulnerable, and survey the building stock. Useful assistance may be found in the following FEMA publications: Rapid Visual Screening of Buildings for Potential Seismic Hazards: A Handbook and Supporting Documentation (FEMA 154 and 155) and the NEHRP Handbook of Techniques for the Seismic Rehabilitation of Existing Buildings (FEMA 172). The Applied Technology Council (ATC) of Redwood City, California, also has available Evaluating the Seismic Resistance of Existing Buildings (ATC-14).

- Request a formal statement on the vulnerability of the types of buildings in the jurisdiction from a qualified structural engineer or organization, university professor, state agency, or consulting structural engineer.

- Secure photographs or slides showing the effects of earthquakes characterized by probable ground motions on buildings like those under consideration. USGS, FEMA, the Earthquake Engineering Research Institute (EERI), and earthquake professionals can provide these.

- Collect data on the building stock and identify the types (structural systems, number of floors, date of construction), numbers, and locations of buildings considered vulnerable. Initially this information may be a general description based on informed judgment.

- Collect property tax assessment data identifying building characteristics, square footage, values, and owner names and addresses.

- Collect occupancy and use information for each building.

- Identify buildings in which hazardous materials are used or stored.

Anticipate uncertainty in expert knowledge as well as disagreements among experts, but work to eliminate the appearance of significant disagreement among credible scientists and engineers by seeking consensus on the most significant points.

Encourage scientists and engineers to debate differences among themselves, ignore minor differences, and publicly air only those disagreements that bear significantly on the policy decisions to be made. Policy-makers with generalist backgrounds should not be expected to resolve technical disagreements, but they can be expected to delay action when seemingly equally qualified scientists and engineers disagree among themselves.

Arising early in Step 1A is the question of the types of buildings considered to be earthquake-vulnerable. Following is a comprehensive list of suspect building types based on earthquake experience and research:

- Unreinforced masonry bearing wall buildings
- Tilt-up concrete wall buildings
- Reinforced masonry wall buildings
- Nonductile concrete moment resisting frame buildings
- Wood frame buildings with soft stories and inadequate foundation connections
- Moment resisting steel frame buildings
- Buildings in areas of expected ground failure
- Earthquake-vulnerable essential buildings

The following profile of typical building uses should be viewed in conjunction with the above list:

- Schools
- Churches
- Hospitals
- Government offices
- Essential services (fire, police, emergency operations, communications, and coordination centers)
- Nonessential services (planning, park and recreation)
- Parking structures
- Residential
- Office/commercial
- Retail
- Manufacturing
- Warehouse
- Industrial
- Public assembly
- Theaters
- Arenas
- Mixed uses

The following outlines various impacts, positive as well as negative, of seismic rehabilitation:

- Lives saved and injuries prevented
- Businesses and homes saved from future damage
- Business and residential disruption prevented
- Increased owner debt and higher loan service payments avoided
- Changed property values and tax levies
- Increased rents
- Some buildings demolished or vacated
- Historic buildings protected
- Other code upgrades triggered (disabled access, energy conservation, asbestos removal, fire sprinkler installation)
- Changed property and other insurance premiums
- Altered availability of loans and insurance

For the affected buildings and neighborhoods, collect data on or at least estimate: the numbers, ages, *income levels, ethnicity, and language capabilities of residents; the numbers and types of businesses and associated employees; the ownership patterns (resident or absent, multiple property and large building owners, government agencies, nonprofit organizations, condominium associations); the property values, loan to equity ratios, mortgage default rates, and rental rates; and the applicable occupancy levels and vacancy rates.*

Evaluate economic data on: the range of costs to rehabilitate typical buildings (for various performance levels) based on structure type, local seismic hazard, and size; the time required to rehabilitate individual building types as well as the whole target set; the potential indirect costs due to the disturbance and displacement caused by the rehabilitation work (lost rent, lost businesses, lost tenants, cost of relocating and inconvenience, and lost sales and property tax revenues); and the future financial benefits of reduced damage.

Many private consulting firms have computer programs and the expertise needed to estimate potential earthquake losses for individual buildings, a portfolio of buildings at different locations, or all buildings within a geographical area. In addition, the National Institute for Building Sciences (NIBS) has released, nonproprietary software ("HAZUS") developed for FEMA that anyone with a desktop computer can use to estimate earthquake losses for their geographic areas.

While data on nationwide earthquake hazards and building stock information from the 1990 census and other data bases will provide at least a general perspective, local information such as that collected as part of this approach can be added and will allow for more accurate planning. Consider using the NIBS software or hiring a firm to use a proprietary program.

Review the results of this preliminary analysis and decide if the seismic risk to the community, company, or owner is significant enough to proceed to the more detailed analysis described in Step 1B.

If the decision is to proceed, prepare a rough estimate of the cost and a schedule to adopt and implement a seismic rehabilitation program.

Step 1B. Detailed Analysis

The information, assumptions, and estimates made in Step 1A should be revisited and additional detail on those points should be sought as part of Step 1B.

Set preliminary earthquake risk reduction objectives: Which buildings? What priorities? What pace? What levels of performance? The following summarizes the performance levels (from greater to lesser) discussed in Chapter 1 of the *Guidelines* and volume:

- Collapse Prevention: means that limiting post-earthquake damage state in which the building is on the verge of experiencing partial or total collapse.

- Life Safety: means that post-earthquake damage state in which significant damage to the structure has occurred, but some margin against either total or partial collapse remains.

- Immediate Occupancy: means that post-earthquake damage state in which only limited structural and non-structural damage has occurred.

- Operational: means that post-earthquake damage state in which the building is suitable for its normal occupancy and use, albeit possibly in a slightly impaired mode.

Performance levels should be matched with building types and functions to determine priorities and pace. In addition, Figure A1 is reproduced here from the *Guidelines* to remind the user of the process for selecting a seismic rehabilitation strategy for a specific building.

Review existing policies, goals, objectives, and requirements in the community to determine how they may "dovetail" or conflict with proposed earthquake risk reduction strategies including land use, economic development, housing, historic preservation, aesthetic and environmental, planned uses for affected areas, future conformance with zoning ordinances, planned changes to infrastructure, compliance with Americans with Disabilities Act (ADA) and other code mandates, compliance with storage and use of hazardous materials regulations, emergency response roles and capabilities, and any other applicable goals, objectives and requirements.

Identify and map hazard areas and affected neighborhoods. Existing maps can be used to identify areas of potential liquefaction and other ground failure as well as areas underlain by soft or saturated soils, including fills over lake and river beds and bay deposits.

Identify neighborhoods or areas where earthquake-vulnerable buildings are highly concentrated.

Consult with the local emergency services manager, fire and police chiefs, and directors of planning, redevelopment, and public works to determine the capability and plans for post-earthquake fire suppression, search and rescue, control of released hazardous materials, damage evaluation, and public safety to see how rehabilitation could reduce post-earthquake demands for their services.

As a collateral benefit, share the information already collected to help these local officials understand their responsibilities and likely problems after an earthquake, use the information derived from these consultations to define problems that can be reduced through seismic rehabilitation, and encourage revision of the emergency response and recovery plans using the information collected.

Identify redevelopment project areas (and funding sources) and consider formation of new projects, possibly expanding the definition of "blight" to include potentially earthquake-vulnerable buildings.

Outline administrative implications including: potential demands for program management (resources and skills); need to support and coordinate proponent activity; need for enhanced enforcement capability (design review and construction inspection); cost of inventories and engineering, economic, social and environmental impact data collection and analysis; cost to support stakeholder participation; cost to implement alternative programs; length of time needed to adopt a program and the approximate duration of the implementation phase; and estimated cost in lost revenues, additional staff requirements, and additional capital outlay to the local government or company.

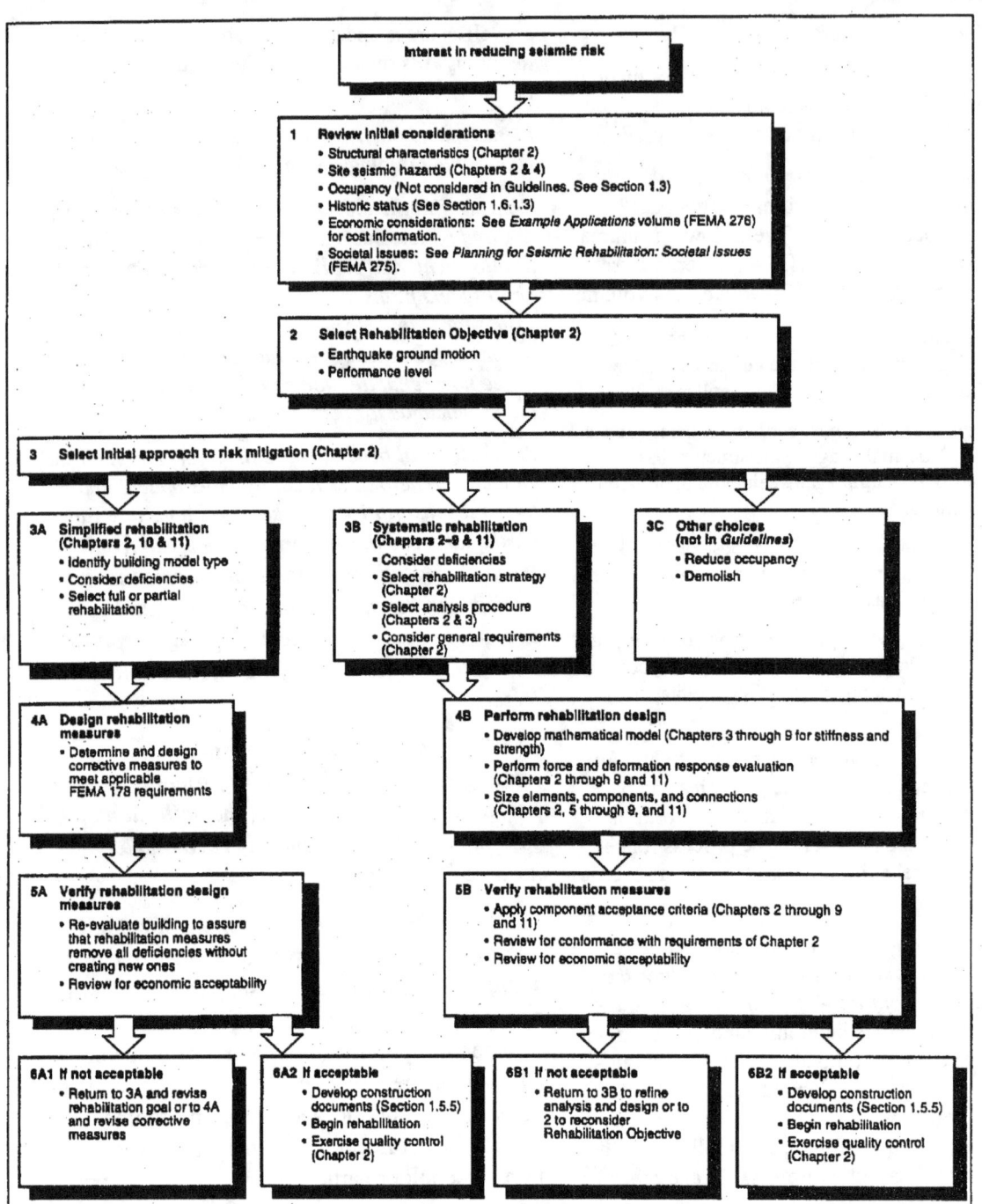

FIGURE A1 Rehabilitation process flowchart
(from Chapter 1, *NEHRP Guidelines for the Seismic Rehabilitation of Buildings*.

Consult legal counsel on the adoption and implementation processes, potential impacts on property rights and leases, and the need to disclose risk information.

Estimate total costs including: cost of engineering and rehabilitation, cost of required other work (ADA compliance, code upgrades), cost of alternative temporary space and relocation, costs of disruption (estimated), possible effect on leases and possible loss of tenants, lost rent and sales during the period of disruption, loss of sales tax revenues, increased debt service for the owner, and increased rent because of the cost of rehabilitation and disruption.

Describe effects that are not quantifiable solely as monetary costs such as loss of housing stock, loss of historical and architecturally important buildings, and business failures, closures and relocations.

Describe trade-off values (amount and cost [direct and indirect]) versus benefits (even if vague, abstract, or probabilistic). The potential bases for justifying seismic rehabilitation include the following:

- Fewer lives lost
- Fewer persons injured
- Less property damage
- Less demand for emergency response
- Less loss of housing resources
- Less loss of historical resources
- Faster economic and social recovery
- Less financial impact of earthquakes
- Less business downtime
- Increased safety for customers/tenants
- Less change for the neighborhood
- Increased building value
- Higher market value for buildings
- Less costly insurance premiums
- More secure equity for loans

Identify existing groups that will be affected by or interested in the seismic rehabilitation program:

- Homeowners associations
- Chambers of commerce
- Merchants associations
- Building and owners managers associations
- Boards of realtors
- Historical and preservation societies
- Ethnic business associations and groups
- Tenant organizations
- Community service clubs
- Labor unions and employee associations

- Civic, religious, fraternal, and other groups

Identify potentially affected autonomous political entities including redevelopment agencies and special districts (fire, police, school, water supply, sanitary, gas, electric and recreation).

Identify expert groups with knowledge to add to the considerations. Some of these groups include:

- Architects
- Civil engineers
- Engineering geologists
- Structural engineers
- Attorneys
- Certified public accountants
- Bankers and financial planners
- Insurers and reinsurers
- University faculties
- Realtors and property managers

Identify those groups directly affected by decisions may not have an effective way to participate in the decision-making process including low income residents of affected buildings, homeless persons, minorities and those with language limitations, elderly and retired persons, and physically challenged persons.

Determine if new organizations are needed to represent previously unorganized groups of affected persons, specific concerns, or issues. If so, identify possible leaders and members to facilitate the formation or representation of the group(s).

Identify potential proponent and opponent leaders, including their respective positions.

Identify news media and meet with reporters and editors to brief them on the concerns and the adoption process, provide background information, and commit to a relationship based on open communication. Media outlets include general circulation daily and weekly papers, ethnic papers, business and legal papers, radio news, television news, and community focused magazines.

Learn how to communicate matters of seismic risk, impacts, conflicting values, and uncertainty to an audience that may not understand the language of science and engineering and may very well have differing values on risk acceptance and the cost of risk reduction.

Accept the idea that people and groups view risk differently and have different values when balancing earthquake risk with other values.

Realize that a mathematical description of risk does not convey a complete message to most people. In addition to describing the probability or chance of an earthquake of a certain magnitude within a year, 30 years or a 100 years, describe what may happen in terms of the damage and the consequences of that damage to a building or the community.

Communicate facts, avoid the temptation to hide impacts or express judgment of others' values, and avoid surprising other participants with information that implies a "hidden agenda."

Deal immediately with concerns raised (even rumors) and solicit expert assistance to address issues and concerns directly.

Provide information on earthquake risk and building vulnerability from trustworthy sources (leaders, officials, expert agencies, professional associations, university faculties).

Provide references where interested parties may obtain more information.

Reconsider loss estimation studies done in Step 1A using new data or, if not done, consider performing these analyses at this point.

Decide whether the seismic risk to the community, company, or owner is significant and whether or not to proceed to Step 2.

Step 2: Develop and Revine Alternatives

Assuming the earthquake hazard and community vulnerability combine to create a seismic risk justifying seismic rehabilitation of certain buildings, Step 2 will result in the definition of practical alternatives. Simply stated, no standard formula or approach will work everywhere. While information already collected may suffice, it often is essential to collect more detailed data (e.g., a property-by-property inventory or consultant analyses of specific issues).

More precise data on the community building stock and its general earthquake-resistance characteristics are almost always needed because many Step 2 dis-

cussions of alternative approaches revolve around the performance levels desired for various types of buildings (and therefore the costs) and the number of buildings potentially involved.

Develop a strategy and a process that will address concerns and involve affected organizations in discussions of alternatives, within the limits posed by available resources and in a reasonable period of time.

Meet with building owners and hear concerns, be open to new or unexpected alternatives, and respect different perceptions.

Provide information to interested individuals and groups on the objectives of possible rehabilitation programs, the seismic hazards, building vulnerability, and the consequences of earthquake damage if nothing is done.

Solicit involvement, comments and suggestions from interested individuals and groups, respond to comments and suggestions, and use informal as well as formal meetings.

Consider formation of an advisory committee and evaluate potential chairs. For the chair, look for a person known for openness and objectivity who is experienced at running meetings, willing to find common ground and build consensus rather than highlight differences and polarize, free from conflict of interest, able to devote the considerable time and energy required, and willing to recommend, support and defend tough decisions and recommendations -- often in public forums.

Regularly meet with and brief council members, corporate decision-makers, or clients on the development of alternatives.

Provide photos of typical and relevant damage and provide documentation of possible damage to the community or company.

Show proof of the seismic hazard.

Describe the possible consequences of likely earthquake damage, both direct (damage to buildings and injuries) and indirect (disruption, loss of tax revenues, loss of housing and historical resources).

Explain the scope and cost of alternative approaches.

Propose an implementation program such as one of the following model programs or a hybrid that combines elements of other models: attrition process, voluntary program, informal/encouragement program, and mandatory program.

Decide which of the building types and uses described above to include.

Decide which neighborhood or geographic areas to include.

Determine if existing plans to upgrade facilities or redevelop an area can be amended to incorporate seismic rehabilitation of buildings.

Decide on a process to enforce the regulations including scopes and deadlines for reports, applications, and work and consider penalties for noncompliance including the possibility of condemnation and demolition.

Reconsider the desired seismic rehabilitation performance levels discussed above according to uses and building types selected in the Step 1A. Decide if it is still feasible to meet those levels in light of the costs, and revisit the performance levels to determine if they are too low to provide the benefits desired or possibly unnecessarily high.

Perform benefit-cost analyses. Because of the difficulty in quantifying the costs and benefits of seismic rehabilitation programs, the low probability of damaging earthquakes and the unpredictability and infrequency but high-consequence of these events, the benefit-cost ratio will often appear unfavorable at first. However, it may not be so when the value of life is taken into account. Nonetheless, the benefit-cost analysis is a good tool to compare alternatives and provides a place to start when considering possibilities to improve the ratio. To this end, consider the following incentives to make seismic rehabilitation less costly and less disruptive to those affected:

- Use preservation tax incentives for historic buildings
- Waive permit and inspection fees
- Waive planning requirements (off-street parking, density restrictions, variance request procedures

- Provide guidance and no-cost inspection services for "do-it-yourself" homeowners
- Allow property tax adjustments and other tax incentives
- Offer loans backed by government bonds
- Form a "Redevelopment Area" and "build-in" seismic rehabilitation
- Use "conservation corps" personnel for some of the work (especially for elderly and low-income residents)
- Increase availability of special purpose construction loans
- Encourage bank/lending institutions to provide incentives
- Secure insurance premium reductions

Solicit comments and advice from the affected parties, their organizations, and the involved professional organizations.

Consider a variety of management solutions that vary with the types of buildings covered by the program (performance objectives, length of time for implementation, triggers, level of building department involvement, incentives).

Decide how long owners should be protected from any new retroactive requirements.

Identify actions to mitigate non-financial impacts of the program.

Determine if and how tenant relocation costs may be funded.

Outline special considerations for historical buildings.

Determine criteria and processes for time extensions.

Revisit the benefits of avoiding future losses, the costs of doing nothing, and the costs of the rehabilitation program selected.

Assess the political feasibility of various options and ask two key questions: Is there enough information and sufficient support to push for action? Is an interim decision or a phased decision-making process appropriate?

Recognize likely pressure to delay action if an earthquake is not perceived as imminent, but recognize pressure to act quickly after an earthquake when repairs and possibilities for rehabilitation are suddenly salient to decision-makers.

Review the strategies available (attrition, voluntary, informal/encouragement, or mandatory) and formulate a recommendation.

Step 3: Adopt an Approach and Implementation Strategy

Once a recommendation to rehabilitate earthquake-vulnerable buildings has been forwarded to the final decision-maker(s), for public agency programs an even more public process begins. A seismic rehabilitation advocate must understand that the decision-maker(s) are expected to request both pro and con information and balance the many needs and capabilities of the community, corporation, or owner. Step 3 uses the results from previous steps to provide the expected information.

Explain the seismic risk and support it with expert testimony.

Determine if seismic rehabilitation can be incorporated into other community programs to improve or redevelop specific areas or facilities.

Explain the benefits, costs, and unquantifiable effects.

Explain the views of those affected.

Explain the reasons for the recommended program in comparison to other possible alternatives.

Anticipate and prepare answers for the following questions: How much will it cost (our city, our company) to comply with the proposed program? How much time do we/I have to make this decision? What is the liability associated with going ahead, or doing nothing? Is there a real earthquake hazard affecting this area? Are standards for seismic rehabilitation available? How can we/I justify imposing this measure (to constituents, a board, a boss, or a client)? What will happen (to the community, business, building or client) if nothing is done? What are neighboring jurisdictions (or competitors) doing?

Recommend and participate in formal hearings.

Modify the recommended program to meet any concerns and to address new information raised during hearings or the formal decision-making process.

Step 4: Secure Resources and Implement

Seismic rehabilitation programs do not run without resources and problems. Their execution requires that resources be committed, processes established, materials prepared, monitoring and evaluations carried out, and adjustments made. Owners of earthquake-vulnerable buildings are seldom well financed, often have difficulty securing new loans, and usually are not experienced in hiring engineers or managing complex construction projects, especially ones that affect other community interests. Step 4 recommends anticipating these conditions.

Obtain funding, qualified staff, office space, equipment, and, if necessary, consultant support.

Prepare and disseminate materials oriented toward all affected parties.

Establish a process for monitoring rehabilitation program progress, identifying problems, and reporting results.

Maintain contact with the organizations and individuals involved with developing the alternatives and adopting the program. Hold meetings with affected groups to facilitate open communications.

Maintain quality control to ensure that projects are properly designed and executed.

In order to protect the credibility of the program, maintain vigilance for over-charging or other fraudulent business practices or incompetent work by engineers, architects, and contractors.

Work with and supply information to building owners to assist them in the wise selection of engineers, architects, and contractors.

Ensure that projects meet requirements to mitigate community impacts.

Be sure that those responsible for offering and managing incentives are responsive to owner needs.

Amend technical provisions of the program whenever the engineering-oriented Guidelines documents are amended.

Be prepared to move quickly if unacceptable or unanticipated side effects occur to avoid creating a political backlash caused by the normal inability to see absolutely every problem ahead of time.

Encourage professional organizations, local colleges, and others to offer training for architects, engineers, plan checkers, inspectors, and construction professionals on following and implementing the Guidelines and their proper execution.

Expect the program to be dynamic and in need of further refinements as a result of experience gained during implementation.

Recommend program refinements to decision-makers when needed.

CONCERNS UNIQUE TO USERS

Depending upon the user (jurisdiction with building code enforcement authority, private or corporate owner, consultant) and the intended application of the *Guidelines*, differing perspectives and problems must be taken into account.

Local Government Building Official Tasks

Design, recommend, advocate, and then implement a seismic rehabilitation program for certain types of building within the jurisdiction. Serve as responsible staff person on the many aspects of the program: seismic risk, engineering, administrative, and possibly even socioeconomic and policy.

Learn what other communities are doing and cooperate to share resources.

Although usually licensed by the state, assess the earthquake engineering capability of local design professionals and contractors to carry out the actual seismic rehabilitation of buildings.

Assess the capability of the building department staff and determine appropriate training needed and its cost.

Self-Motivated Owner Tasks

Recommend to management alternatives for addressing seismic risk.

Locate and engage knowledgeable professionals: geologists and geotechnical engineers, structural engineers, and mechanical/electrical/process engineers.

Consider prior rehabilitation experience and experience using the *Guidelines*.

Consider how to evaluate both single buildings and groups of potentially vulnerable buildings.

Determine the relative importance of various buildings to the company.

Consider building(s) occupancy and functions.

Consider corporate image and reputation with customers and suppliers.

Ensure post-disaster business resumption plans are updated.

Consider post-earthquake access to suppliers, customers, and employees.

Determine geographic distribution of the hazard and the probability of seismic events by region. Quantify the expected seismic loads and determine resulting building vulnerabilities (expected performance under specified loads).

Determine the planning horizon.

Conduct a rapid assessment of buildings.

Determine performance objectives for the company, lines of business and specific facilities.

Do a comparative risk evaluation of facilities considering hazard, vulnerability, and importance.

Determine the seismic rehabilitation requirements, if any, of the jurisdictions responsible for building safety.

Determine availability of external financial incentives.

Determine penalties, if any, for not performing rehabilitation.

Determine if local building or planning regulations will require compliance with other health and safety, access, hazardous material, energy conservation, or historical requirements for each of the buildings found to be vulnerable.

Determine the cost of permits, steps involved, and time requirements to rehabilitate each vulnerable building.

Consider how to benefit from community, customer, and client good will earned by rehabilitating buildings, and determine how to capitalize on these benefits.

Determine if uses and functions at risk are critical, or if redundant facilities provide the necessary back-up at locations outside of the same hazard area.

Determine alternative strategies for meeting desired performance objectives. Have the design consultants do conceptual designs for the following: short-term, temporary measures such as shoring collapse-hazard building elements; nonstructural and falling hazard abatement measures to remove the most vulnerable life-threatening elements; and permanent rehabilitation measures consistent with performance objectives

Identify and meet with persons responsible for the following: operations and business resumption, space management, risk management (including insurance and hazardous materials), emergency response and employee safety, legal counsel, finance, public relations, and government relations.

Survey vacancy rates in nearby buildings to determine the cost and feasibility of temporarily relocating functions during rehabilitation.

Determine knowledge and level of commitment of the upper management and Board of Directors.

Determine responsibility of corporate officers, fiduciary responsibility for the corporation, and personal liability.

Determine the status and flexibility of capital replacement schedules and facility obsolescence.

Review short- and long-term use plans for each building.

Consider competing needs for funds including pressure for short-term profits versus long-term protection of assets, including equipment, buildings, inventory.

Describe the consequences of damage including: business interruption; vulnerability to temporary and permanent loss of market share; reputation for reliability; loss of employees to undamaged competitors; injury to employees; political ramifications, especially if a major local employer or multiple residential or commercial property owner; liability for injuries; off-site consequences of release of hazardous materials; and cost of repairs.

Secure lease or purchase options on alternative space before announcing a need for relocating functions from vulnerable buildings.

Meet with employees and tenants to explain the risk and the steps being taken to address it.

Meet with community groups and local government officials as appropriate.

Evaluate the company's in-house emergency response capability and local government's capability to respond to company problems.

Do a benefit-cost analysis and include a qualitative description of the intangible matters relevant to the decision.

Consulting Design Professional Tasks

Provide professional services to a client seeking to reduce and manage the seismic risk to his or her facilities.

Determine the owner's concerns and objectives and which facilities are involved.

Ask how will priorities be established (risk, occupancy, function, vulnerability, or other factors).

Determine desired performance objectives (which very well may change after risk information and the cost of rehabilitation alternatives are known).

Determine whether risk management measures, (e.g., emergency response and business resumption plans), can be considered as alternatives.

Be certain that the owner understands the possible nonengineering issues, (e.g., relocation, business interruption, costs).

Determine who is responsible for each point under "Self-Motivated Owner" section above.

Secure the engineering and risk management know-how if it does not exist.

Outline any required internal training.

Hire subcontractor specialists.

Determine how knowledge of risk will affect the liability of the firm and client.

Determine how designing to the client's performance objectives using the *Guidelines* will affect your liability.

Appendix B
BSSC SOCIETAL ISSUES PROJECT PARTICIPANTS

PROJECT OVERSIGHT COMMITTEE

Chairman
Eugene Zeller, Director of Planning and Building, Department of Planning and Building, Long Beach, California

ASCE Members
Paul Seaburg, Office of the Associate Dean, College of Engineering and Technology, Omaha, Nebraska
Ashvin Shah, Director of Engineering, American Society of Civil Engineers, Washington, D.C.

ATC Members
Thomas G. Atkinson, Atkinson, Johnson and Spurrier, San Diego, California
Christopher Rojahn, Executive Director, Applied Technology Council, Redwood City, California

BSSC Members
Gerald H. Jones, Consultant, Kansas City, Missouri
James R. Smith, Executive Director, Building Seismic Safety Council, Washington, D.C.

BSSC PROJECT COMMITTEE

Chairman
Warner Howe, Consulting Structural Engineer, Germantown, Tennessee

Members
Gerald H. Jones, Kansas City, Missouri
Harry W. Martin, American Iron and Steel Institute, Auburn, California
Allan R. Porush, Structural Engineer, Dames and Moore, Los Angeles, California
F. Robert Preece, Preece/Goudie and Associates, San Francisco, California
William W. Stewart, FAIA, Stewart·Schaberg/Architects, Clayton, Missouri

Societal Issues Consultant
Robert A. Olson, President, Robert Olson Associates Inc., Sacramento, California

SEISMIC REHABILITATION ADVISORY PANEL

Chairman
Gerald H. Jones, Kansas City, Missouri

Members
David E. Allen, Structures Division, Institute of Research in Construction, National Research Council of Canada, Ottawa, Ontario, Canada
John Battles. Southern Building Code Congress, International, Birmingham, Alabama

David C. Breiholz, Chairman, Existing Buildings Committee, Structural Engineers Association of California, Lomita, California

Michael Caldwell, American Institute of Timber Construction, Englewood, Colorado

Terry Dooley, Morley Construction Company, Santa Monica, California

Steven J. Eder, EQE Engineering Consultants, San Francisco, California

S. K. Ghosh, Mt. Prospect, Illinois

Barry J. Goodno, Professor, School of Civil Engineering, Georgia Institute of Technology, Atlanta

Charles C. Gutberlet, US Army Corps of Engineers, Washington, D.C.

Harry W. Martin, American Iron and Steel Institute, Auburn, California

Margaret Pepin-Donat, National Park Service Retired, Edmonds, Washington

William Petak, Professor, Institute of Safety and Systems Management, University of Southern California, Los Angeles, California

Howard Simpson, Simpson, Gumpertz & Heger, Arlington, Massachusetts

James E. Thomas, Duke Power Company, Charlotte, North Carolina

L. Thomas Tobin, Tobin & Associates, Mill Valley, California

EERI Committee Advisory Committee on Social and Policy Issues

Mary Comerio, University of California, Berkeley

Cynthia Hoover, City of Seattle, Washington

George Mader, Spangle Associates

Robert Olshansky, University of Illinois

Douglas Smits, City of Charleston, South Carolina

Susan Tubbesing, Earthquake Engineering Research Institute

Barbara Zeidman, City of Los Angeles, California

THE COUNCIL: ITS PURPOSE AND ACTIVITIES

Of the National Institute of Building Sciences

The Building Seismic Safety Council (BSSC) was established in 1979 under the auspices of the National Institute of Building Sciences as an entirely new type of instrument for dealing with the complex regulatory, technical, social, and economic issues involved in developing and promulgating building earthquake risk mitigation regulatory provisions that are national in scope. By bringing together in the BSSC all of the needed expertise and all relevant public and private interests, it was believed that issues related to the seismic safety of the built environment could be resolved and jurisdictional problems overcome through authoritative guidance and assistance backed by a broad consensus.

The BSSC is an independent, voluntary membership body representing a wide variety of building community interests (see pages 15-16 for a current membership list). Its fundamental purpose is to enhance public safety by providing a national forum that fosters improved seismic safety provisions for use by the building community in the planning, design, construction, regulation, and utilization of buildings. To fulfill its purpose, the BSSC:

- Promotes the development of seismic safety provisions suitable for use throughout the United States;

- Recommends, encourages, and promotes the adoption of appropriate seismic safety provisions in voluntary standards and model codes;

- Assesses progress in the implementation of such provisions by federal, state, and local regulatory and construction agencies;

- Identifies opportunities for improving seismic safety regulations and practices and encourages public and private organizations to effect such improvements;

- Promotes the development of training and educational courses and materials for use by design professionals, builders, building regulatory officials, elected officials, industry representatives, other members of the building community, and the public;

- Advises government bodies on their programs of research, development, and implementation; and

- Periodically reviews and evaluates research findings, practices, and experience and makes recommendations for incorporation into seismic design practices.

The BSSC's area of interest encompasses all building types, structures, and related facilities and includes explicit consideration and assessment of the social, technical, administrative, political, legal, and economic implications of its deliberations and recommendations. The BSSC believes that the achievement of its purpose is a concern shared by all in the public and private sectors; therefore, its activities are structured to provide all interested entities (i.e., government bodies at all levels, voluntary organizations, business, industry, the design profession, the construction industry, the research community, and the general public) with the opportunity to participate. The BSSC also believes that the regional and local differences in the nature and magnitude of potentially hazardous earthquake events require a flexible approach to seismic safety that allows for consideration of the relative risk, resources, and capabilities of each community.

The BSSC is committed to continued technical improvement of seismic design provisions, assessment of advances in engineering knowledge and design experience, and evaluation of earthquake impacts. It recognizes that appropriate earthquake hazard risk reduction measures and initiatives should be adopted by existing organizations and institutions and incorporated, whenever possible, into their legislation, regulations, practices, rules, codes, relief procedures, and loan requirements so that these measures and initiatives become an integral part of established activities, not additional burdens. Thus, the BSSC itself assumes no standards-making or -promulgating role; rather, it advocates that code- and standards-formulation organizations consider the BSSC's recommendations for inclusion in their documents and standards.

IMPROVING THE SEISMIC SAFETY OF NEW BUILDINGS

The BSSC program directed toward improving the seismic safety of new buildings has been conducted with funding from the Federal Emergency Management Agency (FEMA). It is structured to create and maintain authoritative, technically sound, up-to-date resource documents that can be used by the voluntary standards and model code organizations, the building community, the research community, and the public as the foundation for improved seismic safety design provisions.

The BSSC program began with initiatives taken by the National Science Foundation (NSF). Under an agreement with the National Institute of Standards and Technology (NIST; formerly the National Bureau of Standards), *Tentative Provisions for the Development of Seismic Regulations for Buildings* (referred to here as the *Tentative Provisions*) was prepared by the Applied Technology Council (ATC). The ATC document was described as the product of a "cooperative effort with the design professions, building code interests, and the research community" intended to "...present, in one comprehensive document, the current state of knowledge in the fields of engineering seismology and engineering practice as it pertains to seismic design and construction of buildings." The document, however, included many innovations, and the ATC explained that a careful assessment was needed.

Following the issuance of the *Tentative Provisions* in 1978, NIST released a technical note calling for ". . . systematic analysis of the logic and internal consistency of [the *Tentative Provisions*]" and developed a plan for assessing and implementing seismic design provisions for buildings. This plan called for a thorough review of the *Tentative Provisions* by all interested organizations; the conduct of trial designs to establish the technical validity of the new provisions and to assess their economic impact; the establishment of a mechanism to encourage consideration and adoption of the new provisions by organizations promulgating national standards and model codes; and educational, technical, and administrative assistance to facilitate implementation and enforcement.

During this same period, other significant events occurred. In October 1977, Congress passed the *Earthquake Hazards Reduction Act of 1977* (P.L. 95-124) and, in June 1978, the National Earthquake Hazards Reduction Program (NEHRP) was created. Further, FEMA was established as an independent agency to coordinate all emergency management functions at the federal level. Thus, the future disposition of the *Tentative Provisions* and the 1978 NIST plan shifted to FEMA. The emergence of FEMA as the agency responsible for implementation of P.L. 95-124 (as amended) and the NEHRP also required the creation of a mechanism for obtaining broad public and private consensus on both recommended improved building design and construction regulatory provisions and the means to be used in their promulgation. Following a series of meetings between representatives of the original participants in the NSF-sponsored project on seismic design provisions, FEMA, the American Society of Civil Engineers and the National Institute of Building Sciences (NIBS), the concept of the Building Seismic Safety Council was born. As the concept began to take form, progressively wider public and private participation was sought, culminating in a broadly representative organizing meeting in the spring of 1979, at which time a charter and organizational rules and procedures were thoroughly debated and agreed upon.

The BSSC provided the mechanism or forum needed to encourage consideration and adoption of the new provisions by the relevant organizations. A joint BSSC-NIST committee was formed to conduct the needed review of the *Tentative Provisions*, which resulted in 198 recommendations for changes. Another joint BSSC-

NIST committee developed both the criteria by which the needed trial designs could be evaluated and the specific trial design program plan. Subsequently, a BSSC-NIST Trial Design Overview Committee was created to revise the trial design plan to accommodate a multiphased effort and to refine the *Tentative Provisions*, to the extent practicable, to reflect the recommendations generated during the earlier review.

Trial Designs

Initially, the BSSC trial design effort was to be conducted in two phases and was to include trial designs for 100 new buildings in 11 major cities, but financial limitations required that the program be scaled down. Ultimately, 17 design firms were retained to prepare trial designs for 46 new buildings in 4 cities with medium to high seismic risk (10 in Los Angeles, 4 in Seattle, 6 in Memphis, 6 in Phoenix) and in 5 cities with medium to low seismic risk (3 in Charleston, South Carolina, 4 in Chicago, 3 in Ft. Worth, 7 in New York, and 3 in St. Louis). Alternative designs for six of these buildings also were included.

The firms participating in the trial design program were: ABAM Engineers, Inc.; Alfred Benesch and Company; Allen and Hoshall; Bruce C. Olsen; Datum/Moore Partnership; Ellers, Oakley, Chester, and Rike, Inc.; Enwright Associates, Inc.; Johnson and Nielsen Associates; Klein and Hoffman, Inc.; Magadini-Alagia Associates; Read Jones Christoffersen, Inc.; Robertson, Fowler, and Associates; S. B. Barnes and Associates; Skilling Ward Rogers Barkshire, Inc.; Theiss Engineers, Inc.; Weidlinger Associates; and Wheeler and Gray.

For each of the 52 designs, a set of general specifications was developed, but the responsible design engineering firms were given latitude to ensure that building design parameters were compatible with local construction practice. The designers were not permitted, however, to change the basic structural type even if an alternative structural type would have cost less than the specified type under the early version of the *Provisions*, and this constraint may have prevented some designers from selecting the most economical system.

Each building was designed twice – once according to the amended *Tentative Provisions* and again according to the prevailing local code for the particular location of the design. In this context, basic structural designs (complete enough to assess the cost of the structural portion of the building), partial structural designs (special studies to test specific parameters, provisions, or objectives), partial nonstructural designs (complete enough to assess the cost of the nonstructural portion of the building), and design/construction cost estimates were developed.

This phase of the BSSC program concluded with publication of a draft version of the recommended provisions, the *NEHRP Recommended Provisions for the Development of Seismic Regulations for New Buildings*, an overview of the *Provisions* refinement and trial design efforts, and the design firms' reports.

The 1985 Edition of the *NEHRP Recommended Provisions*

The draft version represented an interim set of provisions pending their balloting by the BSSC member organizations. The first ballot, conducted in accordance with the BSSC Charter, was organized on a chapter-by-chapter basis. As required by BSSC procedures, the ballot provided for four responses: "yes," "yes with reservations," "no," and "abstain." All "yes with reservations" and "no" votes were to be accompanied by an explanation of the reasons for the vote and the "no" votes were to be accompanied by specific suggestions for change if those changes would change the negative vote to an affirmative.

All comments and explanations received with "yes with reservations" and "no" votes were compiled, and proposals for dealing with them were developed for consideration by the Technical Overview Committee and, subsequently, the BSSC Board of Direction. The draft provisions then were revised to reflect the changes deemed appropriate by the BSSC Board and the revision was submitted to the BSSC membership for balloting again.

As a result of this second ballot, virtually the entire provisions document received consensus approval, and a special BSSC Council meeting was held in November 1985 to resolve as many of the remaining issues as possible. The 1985 Edition of the *NEHRP Recommended Provisions* then was transmitted to FEMA for publication in December 1985.

During the next three years, a number of documents were published to support and complement the 1985 *NEHRP Recommended Provisions*. They included a guide to application of the *Provisions* in earthquake-resistant building design, a nontechnical explanation of the *Provisions* for the lay reader, and a handbook for interested members of the building community and others explaining the societal implications of utilizing improved seismic safety provisions and a companion volume of selected readings.

The 1988 Edition

The need for continuing revision of the *Provisions* had been anticipated since the onset of the BSSC program and the effort to update the 1985 Edition for reissuance in 1988 began in January 1986. During the update effort, nine BSSC Technical Committees (TCs) studied issues concerning seismic risk maps, structural design, foundations, concrete, masonry, steel, wood, architectural and mechanical and electrical systems, and regulatory use. The Technical Committees worked under the general direction of a Technical Management Committee (TMC), which was composed of a representative of each TC as well as additional members identified by the BSSC Board to provide balance.

The TCs and TMC worked throughout 1987 to develop specific proposals for changes needed in the 1985 *Provisions*. In December 1987, the Board reviewed these proposals and decided upon a set of 53 for submittal to the BSSC membership for ballot. Approximately half of the proposals reflected new issues while the other half reflected efforts to deal with unresolved 1985 edition issues.

The balloting was conducted on a proposal-by-proposal basis in February-April 1988. Fifty of the proposals on the ballot passed and three failed. All comments and "yes with reservation" and "no" votes received as a result of the ballot were compiled for review by the TMC. Many of the comments could be addressed by making minor editorial adjustments and these were approved by the BSSC Board. Other comments were found to be unpersuasive or in need of further study during the next update cycle (to prepare the 1991 *Provisions*). A number of comments persuaded the TMC and Board that a substantial alteration of some balloted proposals was necessary, and it was decided to submit these matters (11 in all) to the BSSC membership for reballot during June-July 1988. Nine of the eleven reballot proposals passed and two failed.

On the basis of the ballot and reballot results, the 1988 *Provisions* was prepared and transmitted to FEMA for publication in August 1988. A report describing the changes made in the 1985 edition and issues in need of attention in the next update cycle then was prepared. Efforts to update the complementary reports published to support the 1985 edition also were initiated. Ultimately, the following publications were updated to reflect the 1988 Edition and reissued by FEMA: the *Guide to Application of the Provisions*, the handbook discussing societal implications (which was extensively revised and retitled *Seismic Considerations for Communities at Risk*), and several *Seismic Considerations* handbooks (which are described below).

The 1991 Edition

During the effort to produce the 1991 *Provisions*, a Provisions Update Committee (PUC) and 11 Technical Subcommittees addressed seismic hazard maps, structural design criteria and analysis, foundations, cast-in-place and precast concrete structures, masonry structures, steel structures, wood structures, mechanical-electrical systems and building equipment and architectural elements, quality assurance, interface with codes and standards, and composite structures. Their work resulted in 58 substantive and 45 editorial proposals for change to the 1988 *Provisions*.

The PUC approved more than 90 percent of the proposals and, in January 1991, the BSSC Board accepted the PUC-approved proposals for balloting by the BSSC member organizations in April-May 1991.

Following the balloting, the PUC considered the comments received with "yes with reservations" and "no" votes and prepared 21 reballot proposals for consideration by the BSSC member organizations. The reballoting was completed in August 1991 with the approval by the BSSC member organizations of 19 of the reballot proposals.

On the basis of the ballot and reballot results, the 1991 *Provisions* was prepared and transmitted to FEMA for publication in September 1991. Reports describing the changes made in the 1988 Edition and issues in need of attention in the next update cycle then were prepared.

In August 1992, in response to a request from FEMA, the BSSC initiated an effort to continue its structured information dissemination and instruction/training effort aimed at stimulating widespread use of the *NEHRP Recommended Provisions*. The primary objectives of the effort were to bring several of the publications complementing the *Provisions* into conformance with the 1991 Edition in a manner reflecting other related developments (e.g., the fact that all three model codes now include requirements based on the *Provisions*) and to bring instructional course materials currently being used in the BSSC seminar series (described below) into conformance with the 1991 *Provisions*.

The 1994 Edition

The effort to structure the 1994 PUC and its technical subcommittees was initiated in late 1991. By early 1992, 12 Technical Subcommittees (TSs) were established to address seismic hazard mapping, loads and analysis criteria, foundations and geotechnical considerations, cast-in-place and precast concrete structures, masonry structures, steel structures, wood structures, mechanical-electrical systems and building equipment and architectural elements, quality assurance, interface with codes and standards, and composite steel and concrete structures, and base isolation/energy dissipation.

The TSs worked throughout 1992 and 1993 and, at a December 1994 meeting, the PUC voted to forward 52 proposals to the BSSC Board with its recommendation that they be submitted to the BSSC member organizations for balloting. Three proposals not approved by the PUC also were forwarded to the Board because 20 percent of the PUC members present at the meeting voted to do so. Subsequently, an additional proposal to address needed terminology changes also was developed and forwarded to the Board.

The Board subsequently accepted the PUC-approved proposals; it also accepted one of the proposals submitted under the "20 percent" rule but revised the proposal to be balloted as four separate items. The BSSC member organization balloting of the resulting 57 proposals occurred in March-May 1994, with 42 of the 54 voting member organizations submitting their ballots. Fifty-three of the proposals passed, and the ballot results and comments were reviewed by the PUC in July 1994. Twenty substantive changes that would require reballoting were identified. Of the four proposals that failed the ballot, three were withdrawn by the TS chairmen and one was substantially modified and also was accepted for reballoting. The BSSC Board of Direction accepted the PUC recommendations except in one case where it deemed comments to be persuasive and made an additional substantive change to be reballoted by the BSSC member organizations.

The second ballot package composed of 22 changes was considered by the BSSC member organizations in September-October 1994. The PUC then assessed the second ballot results and made its recommendations to the BSSC Board in November. One needed revision identified later was considered by the PUC Executive Committee in December. The final copy of the 1994 Edition of the *Provisions* including a summary of the differences between the 1991 and 1994 Editions was delivered to FEMA in March 1995.

1997 Update Effort

In September 1994, NIBS entered into a contract with FEMA for initiation of the 39-month BSSC 1997 *Provisions* update effort. Late in 1994, the BSSC member organization representatives and alternate representatives and the BSSC Board of Direction were asked to identify individuals to serve on the 1997 PUC and its TSs.

The 1997 PUC was constituted early in 1995, and 12 PUC Technical Subcommittees were established to address design criteria and analysis, foundations and geotechnical considerations, cast-in-place/precast concrete structures, masonry structures, steel structures, wood structures, mechanical-electrical systems and building equipment and architectural elements, quality assurance, interface with codes and standards, composite steel and concrete structures, energy dissipation and base isolation, and nonbuilding structures.

As part of this effort, the BSSC has developed a revised seismic design procedure for use by engineers and architects for inclusion in the 1997 *NEHRP Recommended Provisions*. Unlike the design procedure based on U.S. Geological Survey (USGS) peak acceleration and peak velocity-related acceleration ground motion maps developed in the 1970s and used in earlier editions of the *Provisions*, the new design procedure is based on recently revised USGS spectral response maps. The proposed design procedure involves new design maps based on the USGS spectral response maps and a process specified within the body of the *Provisions*. This task has been conducted with the cooperation of the USGS (under a Memorandum of Understanding signed by the BSSC and USGS) and under the guidance of a five-member Management Committee (MC). A Seismic Design Procedure Group (SDPG) has been responsible for developing the design procedure.

More than 200 individuals have participated in the 1997 update effort, and more than 165 substantive proposals for change have been developed. A series of editorial/organizational changes also have been made. All draft TS, SDPG, and PUC proposals for change were finalized in late February 1997. In early March, the PUC Chairman presented to the BSSC Board of Direction the PUC's recommendations concerning proposals for change to be submitted to the BSSC member organizations for balloting, and the Board accepted these recommendations.

The first round of balloting concluded in early June 1997. Of the 158 items on the official ballot, only 8 did not pass; however, many comments were submitted with "no" and "yes with reservations" votes. These comments were compiled for distribution to the PUC, which met in mid-July to review the comments, receive TS responses to the comments and recommendations for change, and formulate its recommendations concerning what items should be submitted to the BSSC member organizations for a second ballot. The PUC deliberations resulted in the decision to recommend to the BSSC Board that 28 items be included in the second ballot. The PUC Chairman subsequently presented the PUC's recommendations to the Board, which accepted those recommendations.

The second round of balloting was completed on October 27. All but one proposal passed; however, a number of comments on virtually all the proposals were submitted with the ballots and were immediately compiled for consideration by the PUC. The PUC Executive Committee met in December to formulate its recommendations to the Board, and the Board subsequently accepted those recommendations.

The PUC also has identified issues remaining for consideration in the next update cycle and has identified technical issues in need of study. The camera-ready version of the 1997 *NEHRP Recommended Provisions*, including an appendix describing the differences between the 1994 and 1997 edition, was transmitted to FEMA in February 1998. The contract for the 1997 update effort has been extended by FEMA to June 30, 1998, to permit development of a CD-ROM for presentation of the design map data.

Code Resource Development Effort

In mid-1996, FEMA asked the BSSC to initiate an effort to generate a code resource document based on the 1997 Edition of the *Provisions* for use by the International Code Council in adopting seismic provisions for the first edition of the *International Building Code* to be published in 2000.

The orientation meeting of the Code Resource Development Committee (CRDC) appointed to conduct this effort was held in Denver on October 17. At this meeting, the group was briefed on the status of the *Provisions* update effort and formulated a tentative plan and schedule for its efforts.

The group next met in January 1997 to review a preliminary code language/format version of the 1997 *Provisions* and to develop additional needed input. As a result of this meeting, several task groups were established to focus on specific topics and to provide revisions to the preliminary draft. A new draft incorporating these comments then was developed for further refinement by the CRDC. A copy also was delivered to the members of the IBC Structural Subcommittee so that they would begin to have a feeling for where and how the seismic provisions would fit into their code requirements.

The CRDC met again in February to review the second draft of the code language/format version of the 1997 *Provisions*. This meeting was held just preceding a PUC meeting and changes made by the PUC subsequently were incorporated into the CRDC draft. NIBS and CRDC Chairman Gerald Jones presented this composite draft to the IBC Structural Subcommittee on March 1, 1997.

In July, the CRDC met to develop comments on the IBC working draft to be submitted to the ICC in preparation for an August public comment forum. The comments generally reflect actions taken by the PUC in response to comments submitted with the first ballot on the changes proposed for the 1997 *NEHRP Recommended Provisions* as well as CRDC recommendations concerning changes made in the original CRDC submittal by the IBC Structural Subcommittee. CRDC representatives then attended the August forum to support the CRDC recommendations.

The CRDC next met in mid-December to prepare comments on the first published version of the IBC. The proposed "code changes" developed by the committee were submitted to the IBC on January 5, 1998. Subsequent CRDC efforts are expected to focus on supporting the CRDC-developed provisions throughout the code adoption process.

The 2000 Edition

In September 1997, NIBS entered into a contract with FEMA for initiation of the 48-month BSSC effort to update the 1997 *NEHRP Recommended Provisions for Seismic Regulations for New Buildings and Other Structures* for re-issuance in 2000 and prepare code changes based on the 2000 *Provisions* for submittal to the IBC. The BSSC member organization representatives and alternate representatives and the BSSC Board of Direction were asked to identify candidates to participate; the individuals serving on the 1997 update committees were contacted to determine if they are interested in participating in the new effort; and a press release on the 2000 update effort was issued. In addition, the BSSC Board asked 1997 PUC Chair William Holmes of Rutherford and Chekene, San Francisco, if he would be willing to chair the 2000 PUC and he accepted.

In lieu of the Seismic Design Procedure Group (SDPG) used in the 1997 update, the BSSC will re-establish Technical Subcommittee 1, Seismic Design Mapping, used in earlier updates of the *Provisions*. This subcommittee will be composed of an equal number of representatives from the earth science community, including representatives from the USGS, and the engineering community. A sufficient number of members of the SDPG will be included to ensure a smooth transition.

An additional 11 subcommittees will address seismic design and analysis, foundations and geotechnical considerations, cast-in-place and precast concrete structures, masonry structures, steel structures, wood structures, mechanical-electrical systems and building equipment and architectural elements, quality assurance, composite steel and concrete structures, base isolation and energy dissipation, and nonbuilding structures and one ad hoc task group to develop appropriate anchorage requirements for concrete/masonry/wood elements. Unlike earlier updates, it is not anticipated that a technical subcommittee will be appointed to serve as the interface with codes and standards; rather, the PUC will appoint a task group to serve as the liaison with the the model code and standards organizations and three model code representatives will serve on the PUC.

The BSSC, through the PUC and its TS's, will identify major technical issues to be addressed during the 2000 update of the *NEHRP Recommended Provisions*, assess the basis for change to the 1997 Edition, resolve technical issues, and develop proposals for change. The results of recent relevant research and lessons learned from earthquakes occurring prior to and during the duration of the project will be given consideration at all stages of this process. Particular attention will be focused on those technical problems identified but unresolved during the preparation of the 1997 Edition. Attention also will be given to the improvement of criteria to eventually allow for design based on desired building performance levels reflecting the approach taken in the *NEHRP Guidelines for the Seismic Rehabilitation of Buildings*.

The PUC also will coordinate its efforts with those individuals working with the ICC to develop the *IBC*. Changes recommended by those individuals will be submitted to the PUC for consideration and changes developed by the PUC will be formatted for consideration in the *IBC* development process.

As part of the update process, the BSSC also will develop a simplified design procedure in order to improve use of the *Provisions* in areas of low and moderate seismic hazard. This process will be performed by a separate task group reporting directly to TS2, Seismic Design and Analysis.

As in previous update efforts, two rounds of balloting by the BSSC member organizations are planned, and delivery of the final consensus-approved 2000 *Provisions* is expected to occur in December 2000. A report identifying the major differences between the 1997 and the 2000 editions of the *Provisions* and a letter report describing unresolved issues and major technical topics in need of further study also will be prepared.

Following completion of the 2000 *Provisions*, the BSSC will establish a procedure whereby the PUC will prepare code language versions of changes of the *Provisions* for submittal as proposed code changes for the 2003 Edition of the *IBC*. These code changes will be developed for PUC consideration and approval by a Code Liaison Group with the assistance of a consultant experienced in the code change process. In addition, the BSSC will designate three members of the PUC who, along with the consultant, will formally submit the code changes prior to the *IBC* deadline.

Information Dissemination/Technology Transfer

The BSSC continues in its efforts to stimulate widespread use of the *Provisions*. In addition to the issuance of a variety of publications that complement the *Provisions*, over the past seven years the BSSC has developed materials for use in and promoted the conduct of a series of seminars on application of the *Provisions* among relevant professional associations. To date, more than 90 of these seminars have been conducted with a wide variety of cosponsors and more than 70,000 reports have been distributed.

Other information dissemination efforts have involved the participation of BSSC representatives in a wide variety of meetings and conferences, BSSC participation in development of curriculum for a FEMA Emergency Management Institute course on the *Provisions* for structural engineers and other design professionals, issuance of press releases, development of in-depth articles for the publications of relevant groups, work with Building Officials and Code Administrators International (BOCA) that resulted in use of the *Provisions* in the BOCA *National Building Code* and the Southern Building Code Congress International's *Standard Building Code*, and cooperation with the American Society of Civil Engineers (ASCE) that resulted in use of the *Provisions* in the 1993 and 1995 Editions of Standard ASCE 7. In addition, many requests for specific types of information and other forms of technical support are received and responded to monthly.

During 1996, as part of the efforts of a joint committee of the BSSC, Central U.S. Earthquake Consortium, Southern Building Code Congress International and Insurance Institute for Property Loss Reduction to develop mechanisms for the seismic training of building code officials, the BSSC contributed its expertise in the development of a manual for use in such training efforts.

Information dissemination efforts during 1997 have been somewhat curtailed so that resources can be devoted to introduction of the 1997 *Provisions* and related efforts. In this regard, NIBS has requested and received an extension of its existing information dissemination contract with FEMA through September 1998 to permit, among other things, the development of a revised version of a *Nontechnical Explanation of the NEHRP Recommended Provisions* that reflects the 1997 Edition and the structuring of an updated plan to provide informative materials concerning the *Provisions* and the update process.

IMPROVING THE SEISMIC SAFETY OF EXISTING BUILDINGS

Guidelines/Commentary Development Project

In August 1991, NIBS entered into a cooperative agreement with FEMA for a comprehensive 6-year program leading to the development of a set of nationally applicable guidelines for the seismic rehabilitation of existing buildings. Under this agreement, the BSSC serves as program manager with the American Society of Civil Engineers (ASCE) and the Applied Technology Council (ATC) working as subcontractors. Initially, FEMA provided funding for a program definition activity designed to generate the detailed work plan for the overall program. The work plan was completed in April 1992 and in September FEMA contracted with NIBS for the remainder of the effort.

The major objectives of the project were to develop a set of technically sound, nationally applicable guidelines (with commentary) for the seismic rehabilitation of buildings to serve as a primary resource on the seismic rehabilitation of buildings for the use of design professionals, model code and standards organizations, state and local building regulatory personnel, and educators; to develop building community consensus regarding the guidelines; and to develop the basis of a plan for stimulating widespread acceptance and application of the guidelines.

The project work was structured to ensure that the technical guidelines writing effort benefits from: consideration of the results of completed and ongoing technical efforts and research activities as well as societal issues, public policy concerns, and the recommendations presented in an earlier FEMA-funded report on issues identification and resolution; cost data on application of rehabilitation procedures; the reactions of potential users; and consensus review by a broad spectrum of building community interests.

While overall management remained the responsibility of the BSSC, responsibility for conduct of the specific project tasks were shared by the BSSC with ASCE (which organized user workshops and conducted literature review and other research activities) and ATC (which was responsible for drafting the *Guidelines*, its *Commentary*, and a volume of example applications as well as conducting a study to assess the validity of several concepts being proposed for use in the *Guidelines*). Specific BSSC tasks were conducted under the guidance of a BSSC Project Committee. To ensure project continuity and direction, a Project Oversight Committee (POC) was responsible to the BSSC Board for accomplishment of the project objectives and the conduct of project tasks. Further, a Seismic Rehabilitation Advisory Panel was established to review project products and to advise the POC and, if appropriate, the BSSC Board, on the approach being taken, problems arising or anticipated, and progress being made. In addition, three workshops were held over the course of the project to provide the *Guidelines/Commentary* writers with input from potential users of the documents.

The BSSC Board of Direction accepted the 100-percent-complete draft of the *Guidelines* and *Commentary* for consensus balloting in mid-August 1996. The first round of balloting occurred in October-December with a ballot symposium for the voting representatives held in November 1996.

The *Guidelines* and *Commentary* were approved by the BSSC membership; however, a significant number of comments were received. The ATC Senior Technical Committee reviewed these comments in detail and commissioned members of the technical teams that developed the *Guidelines* to develop detailed responses and to formulate any needed proposals for change reflecting the comments. This effort resulted in 48 proposals for change to be submitted to the BSSC member organizations for a second round of balloting.

Following acceptance of the second ballot materials by the BSSC Board, the voting occurred in June-July 1997. Again the results were compiled for review by ATC. Meeting in September 1997, the Project Oversight Committee received recommendations from ATC regarding comment resolution; it was concluded that none of the changes proposed in response to ballot comments were sufficiently substantive to warrant reballoting. Subsequently, the POC conclusion was presented to the BSSC Board, which agreed and approved finalization of the *Guidelines* and *Commentary* for submittal to FEMA for publication. The camera-ready versions of the documents then were prepared and transmitted to FEMA on September 30, 1997.

During the course of the project, BSSC Project Committee recommendations resulted in the following additions to the NIBS/BSSC contract with FEMA for the project: the BSSC ballot symposium for voting representatives mentioned above; the case studies program described below; and an effort to develop the curriculum for and conduct a series of two-day educational seminars to introduce and provide training in use of the *Guidelines* to practicing structural and architectural engineers, seismic engineering educators and students, building officials and technical staff, interested contractors, hazard mitigation officers, and others.

Case Studies Project

The case studies project is an extension of the multiyear project leading to publication of the *NEHRP Guidelines for the Seismic Rehabilitation of Buildings* and its *Commentary* in late 1997. The project is expected to contribute to the credibility of the *Guidelines* by providing potential users with representative real-world application data and to provide FEMA with the information needed to determine whether and when to update the *Guidelines*.

Although the *Guidelines* documents reflect expert experience, current research, and innovative theories, the case studies project is expected to answer a number of critical questions: Can the *Guidelines* and its *Commentary* be understood and applied by practicing design professionals of varying levels of experience? Do the *Guidelines* result in rational designs generated in a reasonable and logical way? What are the costs involved in seismically rehabilitating various types of buildings to the optional levels of performance both above and below the *Guidelines'* "basic safety objective"? Are the requirements to achieve the "basic safety objective" equivalent to, less stringent than, or more stringent than current practice for new construction?

Specifically, the objectives of the project are to: (a) test the usability of the *NEHRP Guidelines for the Seismic Rehabilitation of Buildings* in authentic applications in order to determine the extent to which practicing design engineers and architects find the *Guidelines* documents, including the structural analysis procedures and acceptance criteria, to be presented in understandable language and in a clear, logical fashion that permits valid engineering determinations to be made, and evaluate the ease of transition from current engineering practices to the new concepts presented in the *Guidelines*; (b) assess the technical adequacy of the *Guidelines* design and analysis procedures to determine if application of the procedures results (in the judgment of the designer) in rational designs of building components for corrective rehabilitation measures and whether the designs that result adequately meet the selected performance levels when compared to current practice and in light of the knowledge and experience of the designer; (c) assess whether the *Guidelines* acceptance criteria are properly calibrated to result in component designs that provide permissible values of such key factors as drift, component strength demand, and inelastic deformation at selected performance levels; (d) develop data on the costs of rehabilitation design and construction to meet the *Guidelines'* "basic safety objective" as well as the higher performance levels included and assess whether the anticipated higher costs of advanced engineering analysis result in worthwhile savings compared to the cost of constructing more conservative design solutions arrived at by a less systematic engineering effort; and (e) compare the acceptance criteria of the *Guidelines* with the prevailing seismic design requirements for new buildings in the building location to determine whether requirements for achieving the *Guidelines'* "basic safety objective" are equivalent to or more or less stringent than those expected of new buildings.

It is planned that seismic rehabilitation designs will be developed for over 40 buildings selected insofar as practicable from an inventory of buildings already determined to be seismically deficient under the implementation program of Executive Order 12941 and considered "typical of existing structures located throughout the nation." Where federal buildings from this inventory do not represent the full spectrum of buildings which need to be studied, case study candidates will be sought from among privately owned buildings or those owned by other levels of government. Qualified structural engineering or architectural/engineering (A/E) firms will be engaged to produce detailed designs for seismic rehabilitation of the lateral-load-resisting systems, foundations, and critical nonstructural elements of the selected buildings, and to make specified comparisons with current practices and costs. Each design contractor's products and experiences using the *Guidelines* will be assessed in order to generate credible data that will establish the technical validity of the *Guidelines*, define their economic impact, and identify any needed changes in the *Guidelines* or highlight areas in need of research and investigation before a *Guidelines* update is planned. Many parameters and possible combinations thereof will be considered in addition to basic building types and seismic deficiencies.

The case studies will include consideration of numerous design approaches, options, and determinations to give a balanced representation, within the resources available, of the following factors: different performance levels and ranges, both systematic (linear/nonlinear, static/dynamic) and simplified analysis methods as presented in the *Guidelines*, alternate designs and cost comparisons for the same building provided by more than one design firm, different structural systems, varying seismicity (high, medium, and low), short and stiff versus tall and flexible building types, rehabilitation *Guidelines* compared to current new construction practices, geographic dispersion of cases among seismic risk areas, presence of auxiliary energy dispersion systems or base isolation, and historical preservation status of building.

The project is being guided by the Case Studies Project Committee (CSPC) chaired by Daniel Shapiro, Principal Engineer, SOH and Associates, Structural Engineers, San Francisco, California. The members are: Andrew A. Adelman, P.E., General Manager, Department of Building and Safety, City of Los Angeles, California; John Baals, P.E., Interior Seismic Safety Coordinator, Structural Analysis Group, U.S. Bureau of Reclamation, Denver, Colorado; Jacob Grossman, Principal, Rosenwasser/ Grossman, Consulting Engineers, New York, New York; Edwin T. Huston, Vice President, Smith & Huston, Inc., Seattle, Washington; Col. Guy E. Jester, St. Louis, Missouri; Clarkson W. Pinkham, President, S B Barnes Associates, Los Angeles, California; William W. Stewart, FAIA, Stewart·Schaberg/Architects, Clayton, Missouri; Lowell Shields, Capitol Engineering Consultants, Sacramento, California; Glenn Bell (alternate Andre S. Lamontagne), Simpson, Gumpertz & Heger Inc., Arlington, Massachusetts; Steven C. Sweeney, U.S. Army Construction and Engineering Research Laboratory, Champaign, Illinois.

At its organization meeting in May 1997, the CSPC reviewed the background and structure of the project, developed an initial work plan/project schedule, and defined the roles of the various participants. The CSPC also established three subcommittees to address the development of criteria for building selection, design professional selection, and contractor requests for proposals. In addition to the architects/engineers who will be engaged to perform the case studies designs, the project will utilize a paid Project Technical Advisor and a Design Assessment Panel of professionals knowledgeable about the content and use of the *Guidelines*.

In July, the CSPC met again to review letters of interest and resumes for the advertised position of the Project Technical Advisor; initial selection recommendations were developed for action by the BSSC Board and subsequently resulted in a contract with Andrew T. Merovich of A. T. Merovich and Associates, San Francisco, California. The subcommittee responsible for development of building selection criteria also presented a matrix for the selection and matching of available buildings.

The case studies project was posted in the *Commerce Business Daily* and in the Official Proposals section of *Engineering News Record*. These postings resulted in receipt of 149 expressions of interest; of these, 133 appear to be qualified to move into the next stage of the selection process.

The CSPC is scheduled to meet again on December 2 to finalize the list of buildings recommended for study, approve a draft of the "Request for Qualifications" (RFQ) and contractor selection criteria currently being developed, and identify individuals to serve on the Design Assessment Panel. FEMA has asked that two of the case studies be coordinated with its Disaster Resistant Communities effort by incorporating one building in Seattle, Washington, and one in Oakland, California.

The latest project schedule shows the case study designs being accomplished from May through September 1998 with the final project report to be submitted to FEMA by the end of March 1999.

Earlier Projects Focusing on Evaluation and Rehabilitation Techniques

An earlier FEMA-funded project was designed to provide consensus-backed approval of publications on seismic hazard evaluation and strengthening techniques for existing buildings. This effort involved identifying and resolving major technical issues in two preliminary documents developed for FEMA by others – a handbook for seismic evaluation of existing buildings prepared by the Applied Technology Council (ATC) and a handbook of techniques for rehabilitating existing buildings to resist seismic forces prepared by URS/John A. Blume and Associates (URS/Blume); revising the documents for balloting by the BSSC membership; balloting the documents in accordance with the BSSC Charter; assessing the ballot results; developing proposals to resolve the issues raised; identifying any unresolvable issues; and preparing copies of the documents that reflect the results of the balloting and a summary of changes made and unresolved issues. Basically, this consensus project was directed by the BSSC Board and a 22-member Retrofit of Existing Buildings (REB) Committee composed of individuals representing the needed disciplines and geographical areas and possessing special expertise in the seismic rehabilitation of existing buildings. The consensus approved documents (the *NEHRP Handbook for the Seismic Evaluation of Existing Buildings* and the *NEHRP Handbook of Techniques for the Seismic Rehabilitation of Existing Buildings*) were transmitted to FEMA in mid-1992.

The BSSC also was involved in an even earlier project with the ATC and the Earthquake Engineering Research Institute to develop an action plan for reducing earthquake hazards to existing buildings. The action plan that resulted from this effort prompted FEMA to fund a number of projects, including those described above.

Assessment of the San Francisco Opera House

In October 1994, the NIBS-BSSC initiated an effort to provide FEMA with objective expert advice concerning the San Francisco War Memorial Opera House. The Opera House, constructed circa 1920 with a steel frame clad and infilled with masonry, was damaged in the Loma Prieta earthquake and the city of San Francisco subsequently petitioned FEMA for supplemental funding of approximately $33 million to cover the costs of a complete seismic upgrade of the building under the *Stafford Act*, which provides funding for work when local building code upgrade requirements are met. In this case, the *San Francisco Building Code* was the local code in effect. The effort was structured to involve three phases, if warranted, and was to be conducted by a three-member Independent Review Panel of experts knowledgeable and experienced in building codes and building code administration.

During Phase I, the Review Panel conducted an unbiased, expert review of the applicable code sections pertinent to the repair of earthquake damage in order to provide FEMA with a definitive interpretation of such terms as "how much" change/repair of "what nature" would be sufficient to require complete seismic upgrading of a building of the same general type and construction as the Opera House. It reviewed all relevant, immediately available information about the Opera House case provided by FEMA and the city and the relevant portions of the *San Francisco Building Code* and other similar building codes pertinent to the repair of earthquake-caused damage to buildings and prepared and delivered to FEMA in February 1995 a preliminary report of its findings.

At this point, the Panel was informed by FEMA that the city of San Francisco had rescinded its request indicating that the "proposed determination on eligibility for funding through review and recommendation by an independent and impartial review body from NIBS" would not be necessary. Later, however, FEMA asked that NIBS-BSSC complete Phase I so that it would be better prepared should other similar situations arise. Thus, the Panel continued and delivered a final report to FEMA in July 1995.

IMPROVING THE SEISMIC SAFETY OF NEW AND EXISTING LIFELINES

Given the fact that buildings continue to be useful in a seismic emergency only if the services on which they depend continue to function, the BSSC developed an action plan for the abatement of seismic hazards to lifelines to provide FEMA and other government agencies and private sector organizations with a basis for their long-range planning. The action plan was developed through a consensus process utilizing the special talents of individuals and organizations involved in the planning, design, construction, operation, and regulation of lifeline facilities and systems.

Five lifeline categories were considered: water and sewer facilities, transportation facilities, communication facilities, electric power facilities, and gas and liquid fuel lines. A workshop involving more than 65 participants and the preparation of over 40 issue papers was held. Each lifeline category was addressed by a separate panel and overview groups focused on political, economic, social, legal, regulatory, and seismic risk issues. An Action Plan Committee composed of the chairman of each workshop panel and overview group was appointed to draft the final action plan for review and comment by all workshop participants. The project reports, including the action plan and a definitive six-volume set of workshop proceedings, were transmitted to FEMA in May 1987.

In recognition of both the complexity and importance of lifelines and their susceptibility to disruption as a result of earthquakes and other natural hazards (hurricanes, tornadoes, flooding), FEMA subsequently concluded that the lifeline problem could best be approached through a nationally coordinated and structured program aimed at abating the risk to lifelines from earthquakes as well as other natural hazards. Thus, in 1988, FEMA asked the BSSC's parent institution, the National Institute of Buildings Sciences, to provide expert recommendations concerning appropriate and effective strategies and approaches to use in implementing such a program.

The effort, conducted for NIBS by an ad hoc Panel on Lifelines with the assistance of the BSSC, resulted in a report recommending that the federal government, working through FEMA, structure a nationally coordinated, comprehensive program for mitigating the risk to lifelines from seismic and other natural hazards that focuses on awareness and education, vulnerability assessment, design criteria and standards, regulatory policy, and continuing guidance. Identified were a number of specific actions to be taken during the next three to six years to initiate the program.

MULTIHAZARD ACTIVITIES

Multihazard Assessment Forum

In 1993, FEMA contracted with NIBS for the BSSC to organize and hold a forum intended to explore how best to formulate an integrated approach to mitigating the effects of various natural hazards under the National Earthquake Hazards Reduction Program. More than 50 experts in various disciplines concerning natural hazards risk abatement participated in the June 1994 forum and articulated the benefits of pursuing an integrated approach to natural hazards risk abatement. A BSSC steering committee then developed a report, *An Integrated Approach to Natural Hazards Risk Mitigation*, based on the forum presentations and discussion that urged FEMA to initiate an effort to create a National Multihazard Mitigation Council structured and charged to integrate and coordinate public and private efforts to mitigate the risk from natural hazards. This report was delivered to FEMA in early 1995.

Multihazard Council Program Definition and Initiation

In September 1995, the BSSC negotiated with FEMA a modification of an existing contract to provide for conduct of the first phase of a longer term effort devoted to stimulating the application of technology and experience data in mitigating the risks to buildings posed by multiple natural hazards and development of natural hazard risk mitigation measures and provisions that are national in scope for use by those involved in the planning, design, construction, regulation, and utilization of the built environment. During this first phase, the BSSC is conducting a program definition and initiation effort expected to culminate in the establishment of a National Multihazard Mitigation Council (NMMC) to integrate and coordinate public and private efforts to mitigate the risks associated with natural hazards as recommended in the report cited above.

To conduct the project, the BSSC established a 12-member "blue ribbon" Multihazard Project Steering Committee (MPSC) composed of well-respected leaders in the natural hazards risk mitigation community. The MPSC, which met in July and December 1996 and February 1997, to developed an organizational structure for the proposed council, a draft charter, a draft mission statement, and a preliminary outline for a work plan. Due consideration has been given to the fact that the proposed council will need to maximize the use of resources through mitigation of risks utilizing common measures; promote cost-effective loss reduction, effective technology transfer, conflict identification, and coordination of performance objectives; improve efficiency in the development of codes and standards; provide an open forum for articulation of different needs and perspectives; facilitate policy adoption and implementation; fill educational and public awareness needs; and provide a single credible source for recommendations and directions. In addition, the MPSC is responsible for formulating and directing implementation of a strategy for effectively stimulating the level of interest and degree of cooperation among the various constituencies needed to establish the proposed council.

One of the major project milestones was the organization and conduct of a September 8-10 forum to review the proposed charter, mission statement, and five-year plan. Almost 80 individuals attended. Following background presentations and status reports on current mitigation-related activities, the forum was devoted primarily to presentation and discussion of the preliminary goals and objectives of the proposed council; the proposed NMMC Charter, home/organization, and membership; proposed activities to be included in the five-year plan for the NMMC; and the Steering Committee's candidates for the initial NMMC board. In essence, the forum participants gave consensus approval to the proposed goals, objectives, charter, and membership of the Council and accepted NIBS as the most likely candidate to serve as the home organization of the NMMC.

At its November 1997 meeting, the NIBS Board of Directors reviewed the goals/objectives and activities statements and charter for the NMMC as discussed at the forum. They accepted the charter with some changes. The new council, to be called the Multihazard Mitigation Council (MMC), will now be a sister council to the BSSC and other NIBS councils.

EMI Multihazard Building Design Summer Institute

In 1994, NIBS, at the request of FEMA's Emergency Management Institute (EMI), entered into a contract for BSSC to provide support for the of the EMI Multihazard Building Design Summer Institute (MBDSI) for university and college professors of engineering and architecture. The 1995 MBDSI, conducted in July 1995, consisted of four one-week courses structured to encourage widespread use of mitigation techniques in designing/rehabilitating structures to withstand forces generated by both natural and technological hazards by providing the attending academics with instructional tools for use in creating/updating building design courses.

BSSC MEMBER ORGANIZATIONS

AFL-CIO Building and Construction Trades Department
AISC Marketing, Inc.
American Concrete Institute
American Consulting Engineers Council
American Forest and Paper Association
American Institute of Architects
American Institute of Steel Construction
American Insurance Services Group, Inc.
American Iron and Steel Institute
American Plywood Association
American Society of Civil Engineers
American Society of Civil Engineers--Kansas City Chapter
American Society of Heating, Refrigeration, and Air-Conditioning Engineers
American Society of Mechanical Engineers
American Welding Society
Applied Technology Council
Associated General Contractors of America
Association of Engineering Geologists
Association of Major City Building Officials
Bay Area Structural, Inc.*
Brick Institute of America
Building Officials and Code Administrators International
Building Owners and Managers Association International
Building Technology, Incorporated*
California Geotechnical Engineers Association
California Division of the State Architect, Office of Regulation Services
Canadian National Committee on Earthquake Engineering
Concrete Masonry Association of California and Nevada
Concrete Reinforcing Steel Institute
Earthquake Engineering Research Institute
General Reinsurance Corporation*
Hawaii State Earthquake Advisory Board
Insulating Concrete Form Association
Institute for Business and Home Safety
Interagency Committee on Seismic Safety in Construction
International Conference of Building Officials

International Masonry Institute
Masonry Institute of America
Metal Building Manufacturers Association
National Association of Home Builders
National Concrete Masonry Association
National Conference of States on Building Codes and Standards
National Council of Structural Engineers Associations
National Elevator Industry, Inc.
National Fire Sprinkler Association
National Institute of Building Sciences
National Ready Mixed Concrete Association
Permanent Commission for Structural Safety of Buildings*
Portland Cement Association
Precast/Prestressed Concrete Institute
Rack Manufacturers Institute
Seismic Safety Commission (California)
Southern Building Code Congress International
Southern California Gas Company*
Steel Deck Institute, Inc.
Steel Joist Institute*
Steven Winter Associates, Inc.*
Structural Engineers Association of Arizona
Structural Engineers Association of California
Structural Engineers Association of Central California
Structural Engineers Association of Colorado
Structural Engineers Association of Illinois
Structural Engineers Association of Northern California
Structural Engineers Association of Oregon
Structural Engineers Association of San Diego
Structural Engineers Association of Southern California
Structural Engineers Association of Utah
Structural Engineers Association of Washington
The Masonry Society
U. S. Postal Service*
Western States Clay Products Association
Western States Council Structural Engineers Association
Westinghouse Electric Corporation*
Wire Reinforcement Institute, Inc.

---

* Affiliate (non-voting) members.

(January 1998)

BUILDING SEISMIC SAFETY COUNCIL PUBLICATIONS

Available free from the Federal Emergency Management Agency at 1-800-480-2520
(order by FEMA Publication Number)

For detailed information about the BSSC and its projects, contact:
BSSC, 1090 Vermont Avenue, N.W., Suite 700, Washington, D.C. 20005
Phone 202-289-7800; Fax 202-289-1092; e-mail cheider@nibs.org

NEW BUILDINGS PUBLICATIONS

The NEHRP (National Earthquake Hazards Reduction Program) Recommended Provisions for Seismic Regulations for New Buildings, 1997 Edition, 2 volumes and maps (FEMA Publication 302 and 303)—printed copies expected to be available in early 1998.

The NEHRP (National Earthquake Hazards Reduction Program) Recommended Provisions for Seismic Regulations for New Buildings, 1994 Edition, 2 volumes and maps (FEMA Publications 222A and 223A).

The NEHRP (National Earthquake Hazards Reduction Program) Recommended Provisions for the Development of Seismic Regulations for New Buildings, 1991 Edition, 2 volumes and maps (FEMA Publications 222 and 223) — limited to existing supply.

Guide to Application of the 1991 Edition of the NEHRP Recommended Provisions in Earthquake Resistant Building Design, Revised Edition, 1995 (FEMA Publication 140)

A Nontechnical Explanation of the NEHRP Recommended Provisions, Revised Edition, 1995 (FEMA Publication 99)

Seismic Considerations for Communities at Risk, Revised Edition, 1995 (FEMA Publication 83)

Seismic Considerations: Apartment Buildings, Revised Edition, 1996 (FEMA Publication 152)

Seismic Considerations: Elementary and Secondary Schools, Revised Edition, 1990 (FEMA Publication 149)

Seismic Considerations: Health Care Facilities, Revised Edition, 1990 (FEMA Publication 150)

Seismic Considerations: Hotels and Motels, Revised Edition, 1990 (FEMA Publication 151)

Seismic Considerations: Office Buildings, Revised Edition, 1996 (FEMA Publication 153)

Societal Implications: Selected Readings, 1985 (FEMA Publications 84)

EXISTING BUILDINGS PUBLICATIONS

NEHRP Guidelines for the Seismic Rehabilitation of Buildings, 1997 (FEMA Publication 273)

NEHRP Guidelines for the Seismic Rehabilitation of Buildings: Commentary, 1997 (FEMA Publication 274)

Planning for Seismic Rehabilitation: Societal Issues, 1998 (FEMA Publication 275)

Example Applications of the NEHRP Guidelines for the Seismic Rehabilitation of Buildings, to be available in mid-1998 (FEMA Publication 276)

NEHRP Handbook of Techniques for the Seismic Rehabilitation of Existing Buildings, 1992 (FEMA Publication 172)

NEHRP Handbook for the Seismic Evaluation of Existing Buildings, 1992 (FEMA Publication 178)

An Action Plan for Reducing Earthquake Hazards of Existing Buildings, 1985 (FEMA Publication 90)

MULTIHAZARD PUBLICATIONS

An Integrated Approach to Natural Hazard Risk Mitigation, 1995 (FEMA Publication 261/2-95)

LIFELINES PUBLICATIONS

Abatement of Seismic Hazards to Lifelines: An Action Plan, 1987 (FEMA Publication 142)

Abatement of Seismic Hazards to Lifelines: Proceedings of a Workshop on Development of An Action Plan, 6 volumes:

Papers on Water and Sewer Lifelines, 1987 (FEMA Publication 135)

Papers on Transportation Lifelines, 1987 (FEMA Publication 136)

Papers on Communication Lifelines, 1987 (FEMA Publication 137)

Papers on Power Lifelines, 1987 (FEMA Publication 138)

Papers on Gas and Liquid Fuel Lifelines, 1987 (FEMA Publication 139)

Papers on Political, Economic, Social, Legal, and Regulatory Issues and General Workshop Presentations, 1987 (FEMA Publication 143)